A CULTURAL HISTORY OF CLIMATE

A CULTURAL HISTORY OF CLIMATE

WOLFGANG BEHRINGER

Translated by Patrick Camiller

polity

First published in German as *Kulturgeschichte des Klimas* © Verlag C. H. Beck oHG, München, 2007

This English edition © Polity Press, 2010

The translation of this work was supported by a grant from the Goethe-Institut which is funded by the German Ministry of Foreign Affairs.

Reprinted 2010, 2013 (twice)

Polity Press
65 Bridge Street
Cambridge CB2 1UR, UK

Polity Press
350 Main Street
Malden, MA 02148, USA

ISBN-13: 978-0-7456-4528-5 (hardback)
ISBN-13: 978-0-7456-4529-2 (paperback)

A catalogue record for this book is available from the British Library.

Typeset in 10.5 on 12 pt Sabon by Toppan Best-set Premedia Limited
Printed and bound in Great Britain by the MPG Books Group

The publisher has used its best endeavours to ensure that the URLs for external websites referred to in this book are correct and active at the time of going to press. However, the publisher has no responsibility for the websites and can make no guarantee that a site will remain live or that the content is or will remain appropriate.

Every effort has been made to trace all copyright holders, but if any have been inadvertently overlooked the publisher will be pleased to include any necessary credits in any subsequent reprint or edition.

For further information on Polity, visit our website: www.politybooks.com

The translation of this work was supported by a grant from the Goethe-Institut which is funded by the German Ministry of Foreign Affairs.

CONTENTS

CONTENTS

PREFACE

Repentance preachers blamed the sins of humanity for the climatic vagaries of the Little Ice Age: an immediate change in behaviour would supposedly calm God's wrath and bring about better times. But the weather did not improve, even after scapegoats were identified and hunted down.

Environmental sins are today described as the transgressions that are bringing about manmade climate change. But will an immediate shift in behaviour or a hunt for scapegoats halt the climate change? The answer is: no.

As concerned researchers now realize, scientific analysis alone is not a sufficient answer to the problem. The enforceability of solutions depends on whether they are compatible with cultural images and tendencies in the contemporary world. To understand why this is so, we need not only a pure history of the climate but also a *cultural history*.[1]

The subject of this book is cultural reactions to climate change. In order to get on the trail of the natural mutability of the climate, we shall first take a little journey through the history of the Earth. Then, in a second stage, we shall observe how culture and society have reacted to it.

A major focus will be the only climate crisis that can be well reconstructed from the available sources. Taking the case of the Little Ice Age, we can see what answers were found in one such crisis and consider how they relate to the problems we face today. The Little Ice Age may be regarded as a trial run for *global warming*. We shall learn from it that even minor changes in the climate may result in huge social, political and religious convulsions.

This example also tells us the form in which people coped with the threat of a climate crisis: that is, those in positions of responsibility stopped approaching the crisis with fixed dogmas and gave up looking for moral culprits, sinners or scapegoats. The solutions no longer looked as they had at the beginning of the crisis. They would not be without relevance for ourselves: they led to the world we know today, the 'modern' world after the revolutions in science, communications and agricultural and industrial production. The world did not collapse. Instead, the crisis provoked a flexible cultural response, and even a lasting improvement in living conditions.

In the present situation, climate researchers appear as prophets of an impending catastrophe. We have them to thank for drawing our attention to global warming. But we also remember that for a long time they kept warning us of global cooling and called for measures that now seem absurd. Politicians should have no illusions about the precision of forecasts. What seems true today may tomorrow be discarded as a thing of the past.

Global warming is a serious challenge. The world climate conferences have laid down sensible standards. One can, however, imagine a public discussion that dealt more creatively with the issue. Until now the idea that today's warming holds not only dangers but also possibilities has received scarcely any attention. A cultural history of the climate may serve as a stimulus in that direction.

LIST OF FIGURES

INTRODUCTION

Our investigations begin with three diagrams from the first report of the Intergovernmental Panel on Climate Change (IPCC), in the year 1990. They bring out very clearly the mutability of the climate over the last million years. The top diagram shows the temperature pattern during the Ice Age: the few brief *interglacials*, in which it was warmer than today, appear as precious exceptions and suggest the vulnerability of our modern warm period. The broken line represents the mean value of the normal period from 1961 to 1990. The middle diagram shows the small temperature shifts during the last ten thousand years, since the end of the last great ice age; the climate optimum occurred five to six thousand years ago – that is, in the fourth millennium BC. The bottom diagram shows the temperature pattern during the last thousand years. Here the medieval optimum stands out both against the Little Ice Age and against the signs of a new warming since 1900 – although by 1990 this was still far below the level of the High Middle Ages, not to speak of the maximum temperature in the Holocene.

Many observers found this scenario unsettling. After decades of assuming that the world climate was essentially stable, or even that Gaia – James Lovelock's Earth goddess – would automatically compensate for all disturbances,[1] the presentation of such a degree of volatility came as quite a surprise. Others felt confirmed in their views, since in the 1960s, after a few cooler years, there had been a lively debate about the likelihood of global cooling. As many researchers saw it, however, these graphs played down the most recent and alarming scientific findings. Since the late 1970s it was not global cooling but *global warming* that had been on the agenda, and this aspect did not feature prominently enough in the IPCC graphs of

1

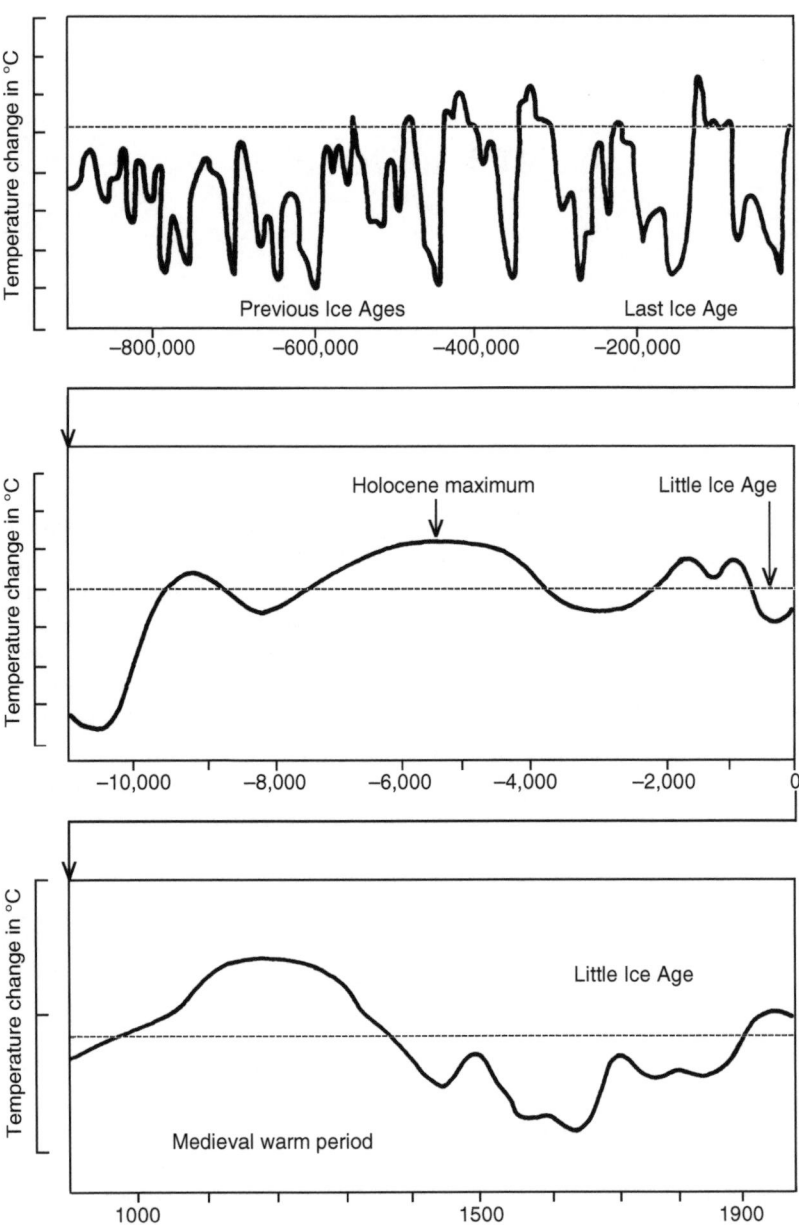

0.1 The fable of climate balance was already disproved in the first IPCC report, in 1990. Whether in the past million years, the past 12,000 years or the past thousand years, we find constant variation between cold and warm periods.

1990. After all the cooling, the graphs even made warming appear desirable, whereas a growing majority of researchers already saw this as a great danger because of associated changes in the composition of the atmosphere. They would therefore try to ensure that the graphic representation of climate history looked quite different in the next major report.

The significance of the hockey stick

The July 2005 issue of *Nature* opened with a report that the Texan Republican congressman Joe Barton (b. 1949), member of the U.S. House Committee on Energy and Commerce, was demanding on behalf of taxpayers an explanation for the research work of three climate specialists. He wanted information about their academic history and funding, as well as access to their data and computer programs.[2] Previously Barton had attacked the same researchers in the *Wall Street Journal* for their allegedly defective scientific methods. Their work had influenced the final report of the Intergovernmental Panel on Climate Change in 2001, which had pilloried the environmental policy of the Bush administration.[3]

The academic world felt that research freedom was in danger. Indeed, since the end of Bill Clinton's presidency (1993–2001), the federal authorities had repeatedly exerted political pressure on scientists.[4] Owing to their uncompromising attitude, climate researchers found themselves in the firing line of party politics. The Californian Democratic congressman Henry Waxman called upon Barton to withdraw his letters. Climate experts received backing from US and international scientific institutions such as the National Science Foundation, the American Association for the Advancement of Science, the president of the National Academy of Sciences, and the European Geosciences Union.

In the eye of this political hurricane were the originators of the 'hockey stick theory': the climate researchers Michael Mann (Pennsylvania State University), Raymond S. Bradley (University of Massachusetts) and Malcolm K. Hughes (University of Arizona), who in 1998 had presented a study on global warming in the past six hundred years. They claimed that on average the 1990s had been hotter than any other decade in the last six centuries, and that this global warming was 'anthropogenous', that is, attributable to greenhouse gases produced by human activity.[5] Their climate curve did not cause much surprise at first, as the longest stretch of the period in question had been marked by the global cooling of the

Little Ice Age. But, shortly before the millennium celebrations, the climate researchers increased its scope by another four centuries. The climate curve of the last thousand years then acquired the form of a hockey stick: not much happened for nine hundred years, but then in the late twentieth century the temperature curve showed a rapid rise. This recent global warming, it was claimed, had no precedent in history. The form of the hockey stick came to symbolize this way of seeing things.[6]

Climate history as a political issue

In 2001 the dispute over the hockey stick curve began to take on an almost religious quality. Its proponents made it their main argument for the signing of the Kyoto Protocol, under which thirty-six industrial countries had undertaken four years earlier to reduce their waste gases. At the time of the dispute there was still a question mark over ratification of the protocol, and only in November 2004 did Russia's final approval remove the last obstacle. In February 2005 the agreements on greenhouse gas reductions and emissions trading came into force for the 2008–2012 period. Of the industrial countries, Australia and the United States (the world's largest emitter of pollutants harmful to the climate) remained outside the treaty.[7] It should be remembered that the Kyoto Protocol had already been agreed, mainly on the basis of the IPCC reports of 1990 and the 1996 update, when the hockey stick curve first appeared on the scene.[8]

The dispute goes on. As each side thinks it is in possession of the truth, it makes opponents out to be paid agents. The coal and oil industry, with which the Bush administration is closely connected, is not keen on costly emission reductions and supports business-friendly scientists with research contracts. Robert F. Kennedy Junior, a nephew of the great president, has spoken of 'a small band of industry-funded charlatans'.[9]

Of course, the charge of corruption is also levelled on the other side, as academic working groups have a vital interest in winning support for their projects and their research institutes. Paul Andrew Mayewski, academic director of the path-breaking Palaeoclimatology drilling programme GISP-2, admits in his presentation that climate researchers are not impartial academics but have personal interests of their own that they promote through followers and pressure groups. It is a question of careers, money and power.[10]

Nevertheless, the idea that climate researchers come to their conclusions for financial reasons scarcely holds water. What is true is

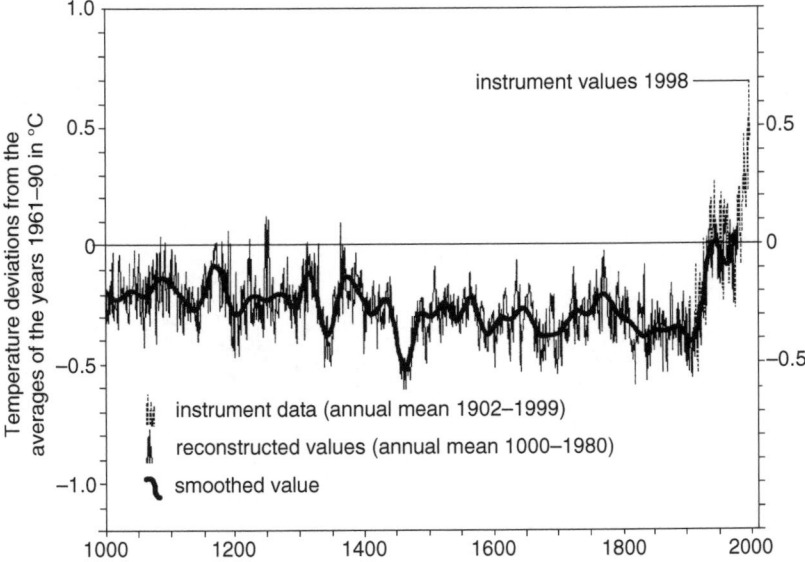

0.2 The climate of the past thousand years as a hockey stick. The IPCC report of 2001 levels out earlier temperature fluctuations by confronting current levels with reconstructed proxy data for previous centuries.

that many of their utterances involve deliberate exaggeration. Thus Stephen H. Schneider (University of California, Stanford), a co-author of the 2001 IPCC report and one of the earliest proponents of the global warming thesis,[11] maintained in an interview: 'To capture the public's imagination . . . we have to offer up scary scenarios, make dramatic, simplified statements.'[12] Yet only a minority of researchers would take such a view.

Apocalyptic scenarios are as counter-productive as the stigmatization of anti-hockeystick critics as 'climate deniers'.[13] In any event, no one denies the existence of the climate, and the question of whether the climate is actually changing has been settled by now. On the second question – whether the change is due only to natural processes or also to human activity and is therefore 'anthropogenous' – there is by now a broad consensus. Reservations are in order only because no one can gaze into the future and because science can scarcely ever achieve absolute certainty. The disgraced prediction of 'global cooling' in the 1960s should serve as a warning. But, with today's evidence and model projections, both the fact of global

Two viewpoints, two sources of profit

0.3 The alarmism of some climatologists strikes many as dubious. Here the thesis of anthropogenic warming is seen as a weapon in the fight for research funding.

warming and its anthropogenous component are already considered 'very probable'.[14] Less clear is how high this component should be set.

Of course, the answer to the 'natural or anthropogenous?' question is of secondary importance if global warming anyway cannot be checked in the short to medium term. But it has played a role in discussion concerning the most appropriate response to climate change. Nowadays climate researchers are more circumspect in dealing with the public.[15] The quasi-religious war of the hockey stick even temporarily reduced public acceptance of the IPCC report and triggered a 'credibility crisis for climatology'. This has not, however, affected the climate change consensus among researchers.[16]

Structure of the book

Whether we are mainly interested in the last few centuries or only in the present or the future, it is necessary to think in terms of long periods. In Chapter 1 some preliminary clarifications are required for an understanding of the later argument. The questions at issue here are: (a) the sources of our knowledge; (b) the mechanisms of natural climate change; and (c) the development of the palaeoclimate from the origins of our planet until the end of the present geological times.

Chapter 2 considers the climate during the epoch of *Homo sapiens sapiens*, from the last great ice age up to the medieval warm period. Unlike the previous chapter, it has a chronological structure. First we consider a time frame of nearly one million years, then discrete periods of several centuries each.

Chapter 3 concerns itself with the symptoms and effects of the Little Ice Age. Climate historians have long held the view that the baneful climate in that period may serve in the construction of broader models.[17] First an attempt is made to grasp the physical and social dimensions of the Little Ice Age, before Chapter 4 tries to get to the bottom of its cultural consequences. These range from the hunt for scapegoats through reflections on human sinfulness to practical adjustments and a successful coming to terms with the crisis by means of the industrialization that led to the world we know today.

Chapter 5 deals with global warming, and includes brief outlines of how the phenomenon was discovered and the debate on its impact. Finally, an Epilogue sums up and weighs some of the results reached in the course of the book.

The endnote references have been kept to a necessary minimum. The literature is cited without subtitles, editions or other precise indications. A few important works are listed in an appendix for special attention and further reading.

— 1 —

WHAT DO WE KNOW ABOUT
THE CLIMATE?

Sources of Climate History

The archive of planet Earth

The expression 'archive of planet Earth' will be used to denote all natural deposits from which scientifically based conclusions may be reached about past climates. The study of this archive has developed enormously since the discovery of radioactivity. The physical basis lies in the fact that the atomic kernels of many elements are unstable, and that they emit radiation as they decay. With the help of mass spectrometers it is thus possible to measure the proportion of mother and daughter elements, and this, together with knowledge of their specific half-value periods, yields evidence as to age. Scientists have also had to deepen their knowledge of the geochemical peculiarities and melting-points of minerals and rocks. By determining the half-value periods of their elements, it is possible to date the age of rocks back to the time of their solidification. This in turn provides information about climate processes.[1]

The *acid isotope method* invented by the US Nobel prizewinner in chemistry Harold C. Urey (1893–1981) has been important for the study of climate history. The discoverer of 'heavy hydrogen' (deuterium) realized that with the help of oxygen atom isotopes it is possible to measure the temperature of the sea in past times. Seawater contains two distinct types of oxygen atoms with different neutron counts: O-18 and O-16. Both are stored in sea organisms, but in a distribution that varies with temperature. The share of heavy acid isotopes (O-18 contains ten instead of eight neutrons) increases in proportion to normal values (O-16) the colder the temperature at the time of depositing.[2] This method revolutionized sedimentation analysis and led to the expansion of deep-sea drilling techniques, with sensational results for ice age research.[3]

A major step forward occurred in the late 1940s when Willard Frank Libby (1908–80) developed the *radiocarbon method* of determining the age of organic remnants ever since the emergence of *Homo sapiens*. This applied both to human skeletons and to many human artifacts. Carbon is stored in plants through photosynthesis, and in animals and humans through respiration. Only with the death of the organism does the replacement process come to a standstill and the process of radioactive disintegration begin. C-14 analysis can help us to identify that point in time. The limits of the radiocarbon method are set by the half-value period of the element C-14 (approx. 5730 +/– 40 years); they lie somewhere between 40,000 and 50,000 years ago.[4]

Sedimentation analysis casts light on the palaeoclimate by providing evidence about climatic conditions (hot or cold, moist or dry), about deposits of plant or animal organisms, about volcanic deposits, sea and lake levels, river terraces, soil horizons and glacier traces. *Palaeobotany* and *palaeozoology* serve to date plant and animal deposits, and in fact the dating of index fossils already has a long tradition going back to the seventeenth century.[5] *Deep-sea drilling technology* has opened up new research opportunities, for the 'ocean memory' offers insights into the development of the soil, the composition of water, the living species and therefore the climate during the periods in question.[6]

Another basic method for the determination of climate is *ice drilling* in the polar caps and the great glaciers, which in the late twentieth century still covered ten per cent of the earth's land surface (against as much as thirty per cent at the height of the last great ice age). In the 1960s the Danish geophysicist Willi Dansgaard (b. 1922) discovered in this a kind of 'time machine' capable of giving fairly precise information about the climate over long periods.[7] Potentially, such analyses can stretch back to the beginnings of the current cool period. Annual deposits can be read from the alternation of dark and light layers in the ice core. Oxygen isotope procedures can then determine the temperature in these deposits, while trapped gas bubbles give direct information about the composition of the air. Furthermore, the radiocarbon method is capable of dating the organic matter in dust particles. Volcanic ash is also present in the dust, and this can be more accurately dated and sorted through the *thermoluminescence* method. Analysis of the

1.1 The main locations of ice bore drilling in Greenland.

sulfate component can provide knowledge about volcanic activity. Years with especially significant eruptions serve as 'pointers' for further calibration of the annual rings.[8]

The ice cores do not only provide multiple and diverse data for evaluation. Their analysis takes us surprisingly far back in time, especially in the case of the earth's two largest ice masses: the Antarctic glaciers in the south polar region, and the Greenland ice sheet near the North Pole. Already in the 1960s, ice cores drilled by the North Greenland Ice Core Project (NGRIP) permitted a more detailed account than scientists had expected of the climate over the past 125,000 years;[9] and the samples taken by GISP2 (Greenland Ice Core Project 2), some thirty kilometres from the site of the first programme, pushed the limit back to 200,000 years.[10] The Russian– French Vostok drilling programme in the Antarctic afforded a view into the past 420,000 years.[11] But the longest retrospect by far came in 2004 with the European Project for Ice Coring in Antarctica

(EPICA). The ice at a depth of 3,270 metres is approximately 800,000 years old and provides a chronicle of the last eight great ice age cycles.[12] This natural 'climate archive' also tells us about the living conditions of primeval humans.[13]

Climate variations during the past 40,000–100,000 years can also be studied through such methods as *varve counting* (analysis of layered deposits in clay sediment, which make it possible to count the years), pollen analysis or *palynology* (analysis of marsh sediment to determine vegetation from the pollen deposits, spores, etc.), *lichenometry* (measurement of regularly growing lichens to study recent glacial peaks and to date terminal moraines),[14] and the science of *dendrochronology* or tree-ring dating (which was first developed in America in the early twentieth century).

The use of tree rings for the study of climate history (*dendroclimatology*) involves dangers, however, which are sometimes underestimated by its proponents. A scanty annual ring does not tell us whether growth was hindered by cold, drought or some other circumstance such as insect attack. And, conversely, an ample ring, which implies favourable growth conditions for a certain type of tree, does not necessarily indicate a good year for crops in general. Grain does not react to wet soil in the same way as oaks or spruces.[15] Unlike climate researchers and archaeologists, interdisciplinary cultural historians are therefore rather sceptical about the significance of annual tree rings for climate studies.[16]

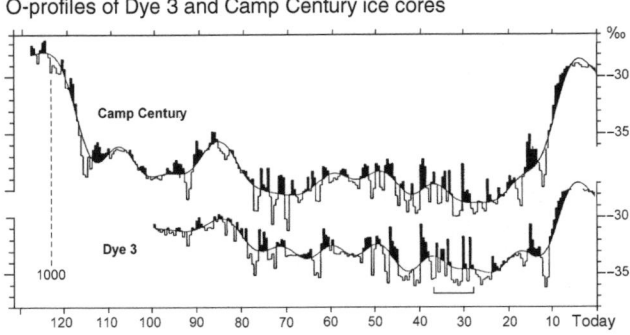

O-profiles of Dye 3 and Camp Century ice cores

1.2 The Danish ice core pioneer, Willi Dansgaard, used oxygen isotope analysis for this new picture of the last great ice age, between the Eem Interglacial and the Holocene.

11

The archive of society

By 'archive of society' we understand deliberate records that are preserved in public or private archives, libraries, data files, and so on. Such archives presuppose recording systems (pictures, numbers, iconograms, characters), and most especially a written language, to perpetuate memory. The archiving of important texts, particularly clay tablets in cuneiform script, began in the ancient civilizations of western Asia; the most important players in this regard were state authorities and religious institutions, which preserved documents, catalogues and correspondence or compiled chronicles (and had inscriptions made on buildings and memorials) to record such events as climatic disasters. Private libraries also entered the picture in Greco-Roman and Chinese antiquity.

In medieval Europe many cities began to keep chronicles, which recorded exceptional weather conditions among other things. The invention of printing in the fifteenth century revolutionized storage and increased the significance of the information market, libraries and the public space.

Among historical records, *weather diaries* stand out as a distinctive text of modern times. Ancient models probably rekindled the love for such books in the European Renaissance, but other factors were the rise of astronomy to become a leading science, and especially an idea of the astronomer Johannes Müller (1436–76), aka Regiomontanus. Müller's astronomical calendar contained not only entries on planetary positions for each day of the year but also an empty line for his own observations. Since astrology saw a direct link between planetary conjunctions and weather conditions, harvests, the availability of wind or water power and other sublunary events, Müller and others tried their hand at predicting the weather and the conjuncture.[17] Users of the almanac would insert notes on the actual weather in the blank lines. The systematic collection of data was supposed to make forecasting more precise, but a comparison between predictions and reality led to sobering results. After more than fifteen years, the Augustinian canon Kilian Leib (1471–1553) critically noted that both peasant wisdom and astrometeorology mostly came up with false predictions. Diaries such as the one kept by Zurich theologian Wolfgang Haller (1525–1601) for the years 1545 to 1576 afford precise insights into daily weather conditions in a critical period of the Little Ice Age.[18]

Numerous records in historical times refer to the quality of the harvest and may be used to obtain what are called *proxy data*. These include direct observation of weather events such as the first snowfall, the duration of snow cover, frozen lakes, rivers and seas, early and late frosts. They also tell us about the growth of plants; in Japan, for ritual reasons, records of the year's first cherry blossoms go back a long way in time. There are data about sowing seasons, fruit blossoms and yields, hay, grain and grape harvests. And lastly we find evidence of harvest quality: the size of the corn tithe (a feudal contribution to the church that was directly correlated with the crop yield);[19] or the prices for bread cereals, which in winter months quite accurately reflected the size of the harvest.[20] The sources further contain many remarks on the quality of produce – of wine, for example, which was usually sour in years with little sun.[21]

Anyone who has worked with late-medieval or early-modern texts will remember continually coming across information of relevance to the climate. Reports of extreme events – drought, flooding, long periods of frost, etc. – can seldom be evaluated on the strength of a single source, since one can never be sure whether they refer to special local conditions, whether the reporter is exaggerating, or whether they are fictions with an allegorical significance. Therefore some climate researchers have collected such local and regional reports in large data banks: the Swiss social historian Christian Pfister for the past five hundred years,[22] or the German geographer Rüdiger Glaser for the past thousand years.[23] The abundance of proxy data from the whole of Europe permits not only statements about the annual climate but at least monthly, and after the year 1500 daily, sometimes hourly, reconstructions of wide-ranging climatic events. On the basis of well-known meteorological correlations, it is even possible to compose retrospective weather maps. What weather forecasts are for meteorologists, 'weather retrocasts' are for climate historians.[24] More generally, works of mythology, literature, art and cartography may also be adduced as climatic evidence, though, as we shall see, not without certain methodological difficulties.[25]

Measuring data with instruments

The calculation of proxy data from weather observations did not become obsolete until long after the appearance of measuring instruments, for the early measurements took place at irregular intervals and used a different scale from that found in later calculations. In 1597 Galileo Galilei designed an instrument for the measurement of air temperature, a *thermometer*. In 1643 his disciple Evangelista Torricelli (1608–47) invented the *barometer* for the measurement of

13

atmospheric pressure. In the 1650s, a first international network was established under Grand Duke Ferdinando de' Medici (1610–70), but the data it produced were difficult to evaluate. There were survey stations in Florence, Bologna, Parma, Milan, Innsbruck, Osnabrück and Paris, and from 1659 in London too. These initiatives were taken up by the Royal Society for the Advancement of Learning, founded in London in 1660, whose first secretary, Robert Hooke (1635–1703), had records kept not only of temperature and pressure but also of wind strength, humidity, cloud cover and the occurrence of fog, rain, hail and snow.[26] At Hooke's suggestion, the Parisian doctor Louis Morin (1635–1715) collected precise daily data over a period of nearly fifty years. The most ambitious international measuring network was the *Societas Meteorologica Palatina*, established in early modern times under the Palatinate Elector Karl Theodor (1724–99). It collected data from as far afield as Spitzbergen and Rome, La Rochelle and Moscow.[27]

Measurement with instruments became more common only in the nineteenth century, when the pioneering role was played by the worldwide British Empire. In the age of Queen Victoria (1819–1901) data were recorded all the way from Europe to India and Australia. Faster means of communication such as the electric telegraph and the underwater cable from Britain to the USA (opened in 1866) increased the range of data collection and evaluation. But the world was still a long way from the instantaneous measurement and satellite transmission with which we have grown familiar since the last third of the twentieth century.

In addition to ground temperature in capital cities and meteorological stations, this period also saw the first measurements of water temperature in the oceans. Since the 1960s, data from weather satellites have provided pictures of the earth for the evening news. And, since the end of the twentieth century, elaborate climate modelling has become possible with a new generation of computers.

Causes of Climate Change

The sun as source of energy

Energy emitted in the sun through nuclear fusion is the basis for chemical, biological and climatic processes on earth. Physicists long thought that the sun's radiated power remained the same (the 'solar constant'), but in reality it undergoes variation. Already in the

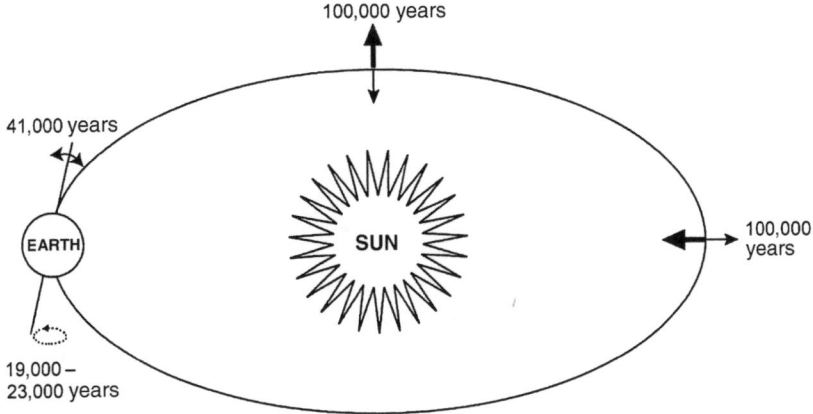

1.3 According to Milankovic, cyclical variations in the earth's orbit explain the periodicity of glaciation cycles.

seventeenth century, telescopes revealed the connection between heat balance and sunspots: a reduction or absence of sunspots usually went together with periods of cooling on earth. Today it is thought possible to establish an eleven-year sunspot cycle, in which periods of reduced solar radiation are caused by, among other things, variations in the earth's orbit.

The Yugoslav astronomer Milutin Milankovic (1879–1958)[28] tried to explain with these variations the more or less regular cycle of cold periods during the Pleistocene. He mentioned, in particular, the earth's eccentric orbit and periodic variations both in the inclination of the earth's axis and in its axis of rotation.[29] The cycles of approximately 100,000 years postulated by Milankovic have been demonstrated in sea drill cores, through the distribution of oxygen isotopes in Pacific plankton. According to the researchers in question, seven such cycles were discovered in roughly the correct position.[30] Ice core researchers also consider the Milankovic cycles relevant for our understanding of ice age formation, and have found evidence of them in ice samples. However, the length of the cycles in the ice over the past 100,000 years does not always correspond precisely to expectations.[31]

The earth's atmosphere

The second factor is the atmosphere around the earth, which determines the effect of solar radiation on the planet. The atmosphere consists of approximately 20 per cent oxygen and nearly 80 per cent nitrogen and trace gases, including 0.03 per cent of the greenhouse gas carbon dioxide (CO_2). Examinations of ice core drilled at the Russian–French Vostok research station in Antarctica have shown that, over the past 420,000 years, the content of trace gases, especially carbon dioxide, in the atmosphere has been directly related to temperatures.[32] A reduced proportion of CO_2 correlates with cooling, an increased proportion with warming. A similar link should be assumed to hold for the preceding period too.

The effect of the earth's atmosphere on the climate strictly corresponds to the *law of the conservation of energy*: solar radiation reaching the earth minus reflected radiation equals the heat radiation from the earth. The oceans and the atmosphere share the heat within the climate system and have an impact on regional climate. The amount of radiation is influenced by the proportion of absorption gases contained in the atmosphere. Apart from water vapour, these gases include trace gases such as methane (CH_4), chlorofluorocarbon (CFC), nitrogen dioxide (N_2O) and carbon dioxide (CO_2). Of the

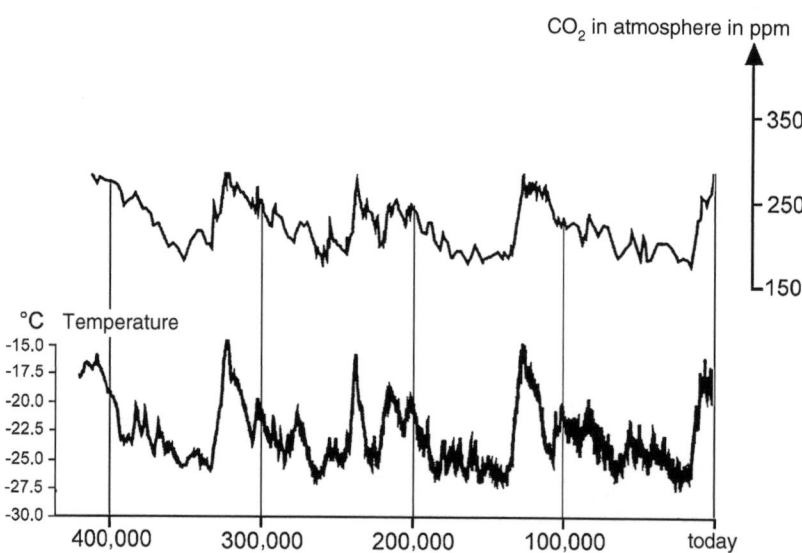

1.4 Evolution of temperature and atmospheric carbon dioxide over the past 420,000 years, according to Vostok ice core drilling in the Antarctic. Evidently there is some connection between the two, but what is the exact causal relationship?

trace gases, carbon dioxide is present in especially variable proportions: 230 ppm (parts per million) at the end of the nineteenth century, but up to 350 ppm at the end of the twentieth. Warm periods in the history of the earth – for example, the Cretaceous (145–65 million years ago), during which dinosaurs ruled the planet – had a proportion of CO_2 as high as 1000 ppm or more. The figure then kept falling, until it reached the low point of the contemporary ice age.[33] For many researchers, the direct correlation between CO_2 content in the atmosphere and global average temperature is a proven fact; it is something like a dogma of contemporary climatology. Yet a comparison from the Vostok ice core shows that the temperature curve and the carbon dioxide curve do not simply run parallel to each other, but that considerable variations may be observed in both amplitude and temporal sequence. Cautious interpreters therefore ask us to remember that 'it is unclear whether changed temperatures are the cause of the changed CO_2 concentrations or vice versa – or whether both may not perhaps be governed by a third, unknown process'.[34]

Plate tectonics

The third factor in the genesis of ice ages is plate tectonics, the shifts in parts of the crust in the earth's upper mantle. Early continental drift over the earth's surface led to changes in ocean currents and – when continents collided with one another and threw up mountain ranges – to changes in wind direction and precipitation patterns. These processes also affected sea levels and the ratio of land to water area. As soon as land masses approach the poles, obstructing the free flow of seawater at these coldest places on earth, ice is formed as a result. The snow and ice cover leads to what is known as the *albedo effect*: more sunlight is reflected, which in turn leads to positive feedback and hence to increased cooling. The share of solar radiation reflected back into space, the so-called *albedo*, reaches nearly 95 per cent above fresh snow, whereas it is less than 10 per cent over the sea. Altogether the albedo is 30 per cent in today's climate system, but it was considerably higher during the ice ages.

According to Steven Stanley, each of the great mass extinctions was caused when a large land mass drifted to one of the polar caps and triggered a lasting formation of ice. Through the feedback effects described above, the planet then sank into a long ice age, which wiped out tropical flora and fauna, in particular, because they had no means of migration. Global icing also changed sea levels: the sea retreated

17

far from the land and altered the shape of coastal regions and conti-nents.[35] In recent geological time, too, continental drift has had sig-nificant effects: five million years ago, for example, the collision of Africa and Eurasia set up the equatorial currents, and the Alps began their folding upwards.[36] The next great event in plate tectonics was the formation of the Central American land bridge, 3.5 million years ago. The closing of the gap between North and South America diverted equatorial currents and led to the appearance of the Gulf Stream, which conveys heat and moisture towards Europe.[37]

Volcanism

The activity of volcanoes is also bound up with plate tectonics. Great eruptions carry ash, aerosol and gases to great heights. Explosive volcanism may also lead to sudden global cooling, if particles of matter are conveyed in large quantity into the stratosphere and driven around the globe by strong winds – which happened in 1815, for example, after the devastating eruption of Tambora in Indonesia.[38] In recent decades there has been some debate about which factors in the volcanic eruptions of the past actually impacted on the global climate. At first, solids in the stratosphere were deemed responsible for the cooling effect, on the grounds that they filter sunlight and were actually observed by people alive at the time. Only in 1963, when height measurements were made after the eruption of Gunung Agung in Bali, was it found that gases are an equally powerful filter and that sulphur plays the decisive role. The 'volcanic explosivity index' (VEI) makes it easier to compare prehistoric eruptions with more recent cases where eye-witnesses have left a first-hand account, such as that of Pliny the Elder (AD 23–79) after the eruption of Vesuvius in AD 79.

For an eruption to affect the climate over great distances, it is nec-essary that large quantities of gases and particles should be thrust up into the stratosphere. This has not been true of most instances in the past ten thousand years, which either offered no more than an inter-esting natural spectacle or else, as in most Icelandic eruptions, devas-tated only the nearby surroundings. Simkin and Siebert have recorded more than five thousand such eruptions for the Holocene period. A global effect on the climate is possible only for 'Plinian' eruptions or for higher than VEI-3 explosions (Grade 3 on the VEI scale).

The mightiest 'ultra-Plinian' eruptions of the past ten thousand years are classified as VEI-7 and 'colossal'. The most recent in this category was the eruption of Tambora in 1815 on the Sunda isles,

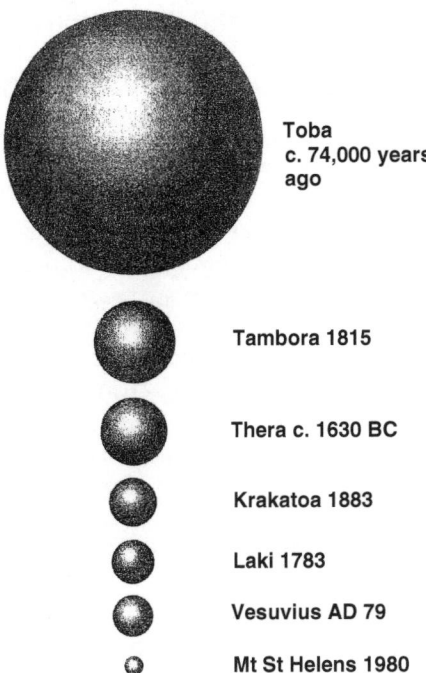

Toba
c. 74,000 years
ago

Tambora 1815

Thera c. 1630 BC

Krakatoa 1883

Laki 1783

Vesuvius AD 79

Mt St Helens 1980

1.5 Volcanic eruptions can produce worldwide cooling. Did the Toba eruption lead to a 'volcanic winter'?

which caused several years of cooling, harvest failure and famine. The fact that there have been no greater explosions (VEI-8 and higher) in the past ten thousand years does not tell us anything, however, about the more distant past.[39] The explosion of Toba in Sumatra some 75,000 years ago probably led to many years of global cooling. This phenomenon, known as 'volcanic winter', contributed to a major round of extinctions.[40]

Meteorites

The conception of large meteorite or asteroid impacts as the causes of doomsday scenarios or mass extinctions enjoys a certain popularity. But serious palaeontologists do not share this enthusiasm. Theoretically, meteorites have similar effects to those of volcanic eruptions, and we shall discuss them in connection with the five mass extinctions of the Phanerozoic.

19

The Palaeoclimate since the Formation of the Planet

Warm periods as the earth's characteristic climate

We are living in an ice age. Many will be surprised to hear this, given all the discussion of global warming. It is therefore necessary, before dealing with the present day, to inform ourselves about the climate during older periods in the earth's history: the *palaeoclimate*. Geology defines an ice age by the existence of glaciers in the polar regions and the high mountains. There have been such ages only five times in the history of our planet: two were in the Precambrian, and two in the Palaeozoic (the oldest part of the Phanerozoic eon, the 'time of visible life'); the fifth is in the Quaternary or – to use the new terminology – the Neocene, the era in which we live today. So, even though it is currently growing hotter, we are still in an ice age. This is very much an exception, for there has been no permanent ice during 95 per cent of the earth's history. Statistically, warm periods are the characteristic climate of our planet – that is, periods in which it has been much hotter than today.

Statements about the climate become more uncertain the further back we go in earth history. According to astrophysics, there was a 'big bang' some 14 billion years ago, out of which the galaxies were formed (11 billion years ago) and then the 'primeval cloud' of our solar system (9 billion years ago). A little less than 5 billion years ago this 'solar cloud' began to collapse, and that was the birth of our

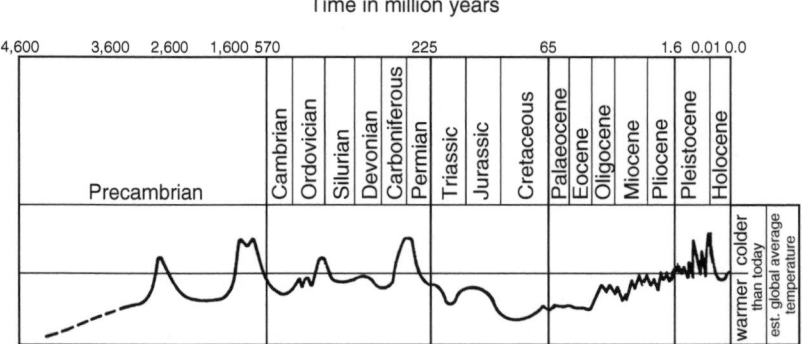

1.6 The palaeoclimate since the earth's formation. It was mostly much warmer than today, but there have been five glacial periods. We are living in one of them.

central star and its planets. The formation of the earth opened the history of its climate. And those beginnings were hot, even infernally hot – which is why the first age of our planet is sometimes called the Hadean Eon, after the Greek word for hell, 'Hades'.[41]

Geologists divide the earth's history into a hierarchical sequence of eons, eras, periods, epochs and ages.[42] The four eons (Hadean, Archean, Proterozoic and Phanerozoic) owe their names to the conditions that they offered for life on the planet. In the formative phase (Hadean) there was still no atmosphere, and so the prerequisites for life on earth were absent. Because of geothermal activity in the formation of the planet, temperatures were higher than they would ever be later in the earth's history. They fell below 100 degrees Celsius some 4 billion years ago, with the formation of the earth's crust. Only then could water condense, and rain, rivers, lakes and seas appear. The oldest known sediments are from 3.7 billion years ago.[43]

In the Archean Eon (approx. 3.8–2.5 billion years), the *primeval atmosphere* took shape on the basis of geophysical processes, possibly volcanic exhalations. Its high CO_2 content ensured strong absorption and low backward dispersion of sunlight. The greenhouse effect promoted favourable temperatures for the first forms of life to appear with *archaeobacteria*. Primeval oceans and continents formed after the condensation of steam. The oldest evidence of a water cycle is found in rocks from 3.2 billion years ago. A large part of the carbon dioxide in the atmosphere disappeared with the water vapour. Roughly 2.6 billion years ago *cyanobacteria* (formerly called blue-green algae) produced oxygen through photosynthesis. The atmosphere, heavy with carbon dioxide, collapsed, and *anaerobic* organisms died out: this was the first mass extinction in earth history. The end of the greenhouse effect brought about global cooling some 2 billion years ago, in the Huronic or Archaic Ice Age.[44]

During the subsequent Proterozoic (approx. 2.5–0.5 billion years), it was for a billion years once again much hotter than in most later periods of earth history. This may have been a result of the new greenhouse effect, set in train by the presence of oxygen in the atmosphere. The oldest plants with a cell nucleus developed during these times, protozoa around 1.4 billion years ago, then finally soft-bodied multicellulars.[45] The climate can be calculated from the oldest fossils, until another severe cooling occurred at the end of this third eon and led to the coldest period in the earth's history. When the primeval continent Rodinia broke up, all the land masses gathered in the equatorial region. Cliffs devoid of vegetation added to the albedo effect. The temperature drop was intensified by ice formation in the polar

21

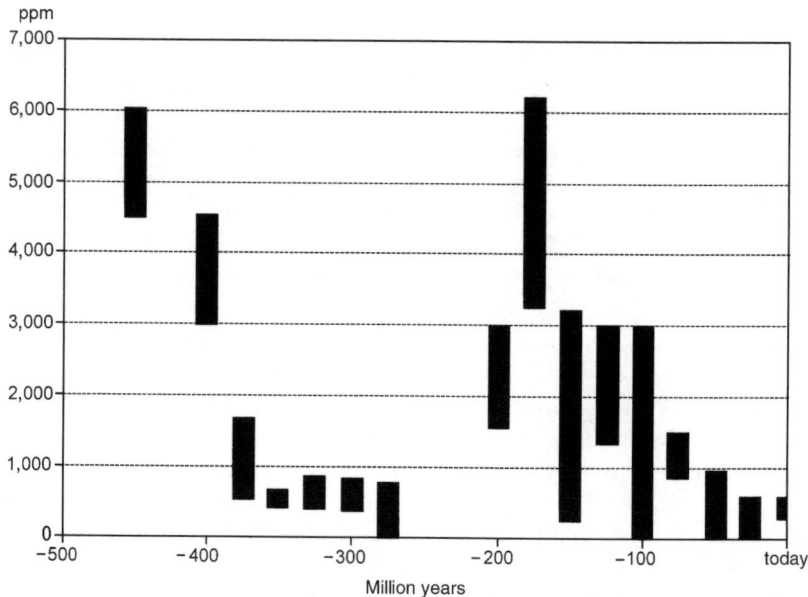

1.7 Another blow against the fable of a constant climate. Before the onset of global warming, CO_2 concentrations were not constant but varied widely over the past 500 million years, even without human influences.

regions,[46] until the icing over of the planet must have made it look from outside like a 'snowball earth'. Under these conditions, the second mass extinction occurred some 650 million years ago. Once more, life on earth would have been close to an end.[47]

The five mass extinctions of the Phanerozoic

The five mass extinctions often mentioned in the literature ('the Big Five') all date from the Phanerozoic. This eon of complex life forms covers only the remaining ten per cent of earth history up to the present day. It is subdivided into the 'eras' of the Palaeozoic (beginning approx. 600 million years ago),[48] the Mesozoic (the age of the dinosaurs, beginning some 250 million years ago)[49] and the Cenozoic (the age of mammals, beginning some 65 million years ago).[50]

We do not know exactly why life on earth got another chance at the end of the early Protozeroic Ice Age. Volcanic eruptions may have freed the earth from its icy girdle by discharging large quantities of CO_2 into the atmosphere and triggering a new greenhouse effect. This would have made life possible again. Some 570 million years ago, during the Cambrian period (600–510 million years), began the explosion of life forms that ushered in the Phanerozoic.[51] In the Cambrian period, with its wealth of fossils (crustaceans, sponges, crabs, snails and eventually the first vertebrates), there are no indications of ice at the polar caps. The climate was globally very hot and remained warmer than today for some 400 million years. In the Ordovician period (510–438 million years), when the number of species multiplied, the most prevalent forms of life were nautiloids, predecessors of the cuttlefish and octopus. But this efflorescence was cut short by the early Ordovician Ice Age, roughly 438 million years ago, in which a large percentage of species died out. In all likelihood this happened because the Gondwana supercontinent, comprising the land masses of the later South America, Africa, Arabia, India, Australia and Antarctica, crossed the South Pole.

The five mass extinctions in the Phanerozoic are each associated with a global cooling:

(1) at the end of the Ordovician, in the 'Ordovician Ice Age';[52]
(2) at the end of the Devonian, as a result of global cooling in the 'late Devonian crisis';[53]
(3) at the end of the Permian, in the 'Permian catastrophe';[54]
(4) at the end of the Triassic;[55]
(5) at the end of the Cretaceous.[56]

At the end of the Permian period, Gondwana and the other land masses grew into the Pangaea supercontinent, which stretched from the South to the North Pole. Thus, the Palaeozoic ended with a phase of terrible cooling, the so-called Permian Ice Age, roughly 250 million years ago. In the space of 10 million years, this 'most horrific of all mass extinctions' consumed 75–95 per cent of species in the marine habitat and – unlike in the previous ice age – a large part of those living on land.[57] The Permian catastrophe is therefore known in the literature as 'the mother of all natural disasters'.[58] It serves as a hiatus between the Palaeozoic and the Mesozoic.

The maximum temperature was probably reached about a hundred million years ago, in the Cretaceous period (146–65 million years), when CO_2 levels in the atmosphere also peaked. The poles acquired roughly their present positions. The polar ice caps melted completely, sea levels rose and the continental shelf was inundated with so-called transgressions. The Alps and the Rocky Mountains soared up as a result of plate tectonics. Many deposits of oil and natural gas were also laid down at that time. On land the first warm-blooded animals, the first mammals and primates, the first primeval hoofed animals and the first birds made their appearance. Dinosaurs roamed northward to Alaska. At the end of the Mesozoic there was another mass extinction, to which the dinosaurs fell victim.[59]

The extinction of the dinosaurs has been blamed on all possible disasters.[60] The drama associated with the asteroid or meteorite version, with the resulting global winter, has made it a special favourite among the public. Such ideas have also been directly inspired by the so-called 'nuclear winter' scenario, the threat of which was widely discussed in the 1980s. The force of the impact is supposed to have triggered flood waves and firestorms and raised so much matter into the stratosphere that the sun was blotted out for months or even years.[61]

Nevertheless, no conclusive proof has yet been adduced for the meteorite theory. Climate disaster caused by a supervolcanic explosion or some unknown cause such as cosmic radiation or disease is another possibility. Nor is it even necessary to think in terms of an exogenous catastrophe. As at the end of the Palaeozoic, climate change at the end of the Mesozoic is fully adequate as an explanation. Temperature changes are the first cause that should come into consideration, because their effect is global and no form of life can escape their influence.[62]

Antarctica as pacemaker for cooling in the Cenozoic

The Cenozoic began 65 million years ago with a primeval catastrophe that brought dramatic changes in the flora and fauna. The Tertiary (or, as it is now called, the Palaeogene) inaugurated the ice age in which we still find ourselves today (Cenozoic Ice Age).[63] The South Pole is presently occupied by Antarctica, a land mass which, because of its permanent glaciation, is not commonly thought of as a continent. From its earlier position close to the Equator – as part of the southern 'Gondwana' continent – this land mass drifted to the southern polar region and, at the start of the Palaeocene (65–55 million years), began to be covered with an ice cap. Cooling was favoured by the appearance of high mountains, wind circulation and precipitation: the Alps, the Rockies and the Cordilleras continued to grow, and some 45 million years ago the Himalayas folded upwards after the collision of India with Asia.[64] Since the separation of Antarctica from Australia in the Oligocene (34–23 million years), the South Pole continent has been surrounded by a cold sea current that blocks warmer currents. Ocean currents still familiar today took shape with

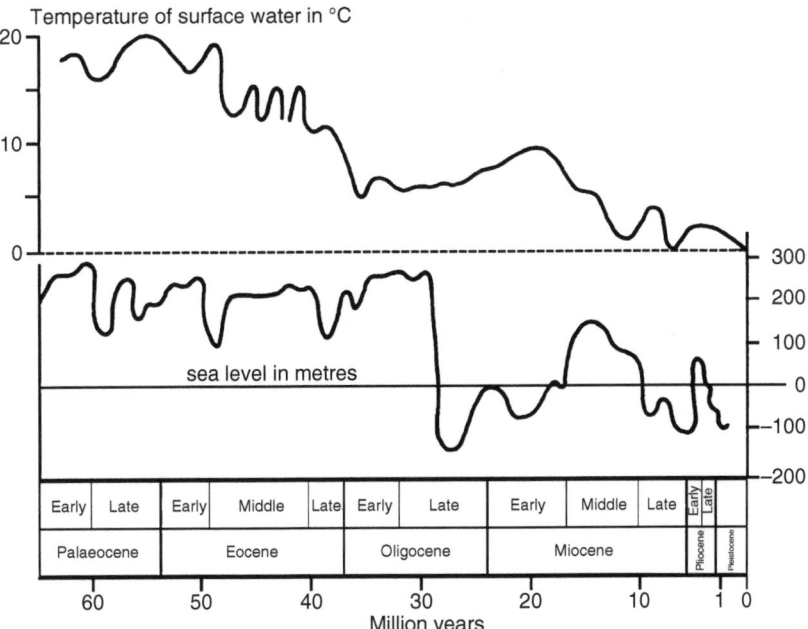

1.8 Variations in sea level and surface temperatures over the past 65 million years. The sharply falling sea level clearly reflects the great ice ages of the Oligocene, the Miocene and the Pleistocene.

the arrangement of the continents in more or less their present form. In the Eocene/Palaeogene there began a sharp cooling that lowered the average annual temperature in Europe from more than 20 degrees Celsius to approximately 12 degrees. The warm-humid climate of the Mesozoic yielded globally to a climate with wet and dry periods and sharp variations of temperature.[65]

The most recent ice age has taken dramatic forms only during the last two million years, in the Pleistocene epoch. From the point of view of human history, this is 'the Ice Age' in the narrow sense of the term, in which not only the Antarctic but also large parts of the northern hemisphere have to endure permanent ice. But the scale of the glaciation has been subject to variations. With the help of oxygen isotope methods, deep sea drilling has already been able to establish quite accurately the course of the climate during the Early Pleistocene (approx. 2.4/1.8 to 0.78 million years). Cold and warm periods alternated with each other in long cycles. With the help of the 'archive of nature' it is possible to identify more than twenty cold and warm periods. The annual average temperature of the air during the cold periods fell as low as 12 degrees Celsius, and ocean surface temperatures to 7 degrees. The snow line in the Alps fell to 1500 metres lower than before, and water binding in glaciers caused global sea levels to fall to around 200 metres lower than before.[66]

These findings for the Early Pleistocene have been confirmed by palaeontological investigation. In the northern hemisphere, continuous cooling was often associated with lower levels of precipitation and corresponding changes in the flora and fauna. In Eurasia and North America, the forests were replaced by forest steppe and grassland, while arid zones covered much of today's African and American savannah and Asiatic forest and grass steppe. Animal migration accompanied the shifting vegetation, as many animals are tied to certain plants for food. In highly specialized cases – amphibians, for example – their natural habitat was reduced. Other animals could survive despite sharp changes in the limits of vegetation, or else – in the case of large mammals such as elephants, rhinoceroses, horses and cattle – profited from the increased pastureland resulting from climatic change. The mammoth, the woolly rhinoceros and the musk ox, all adapted to extreme cold, have been shown to have existed in Central Europe after the end of the Early Pleistocene.[67]

Climate change and human evolution

In a broad view of things, the evolution of man's predecessors was influenced or even determined by geological changes and climatic variations.[68] The formation of the East African Rift Valley through the drift of tectonic plates played a special role in this respect. Great ecological diversity developed in this valley, dominated as it was by climatic instability and constant change. This external insecurity was an engine of evolution. Man's predecessors began to develop in the East African highlands after the massive cooling some ten million years ago. *Australopithecus Africanus* did not come down from the trees but was a species of primate living on the ground. Its habitat increased with the retreat of the forest and the spread of the savannah. Its upright gait helped it to see better in grassland and released its front extremities.[69]

Researchers today do not believe that *Australopithecus* was a 'predator ape', since it lacked the claws and fangs necessary for fighting prey. It probably lived on the carrion that the abundant megafauna left behind. These early humans offset with speed and

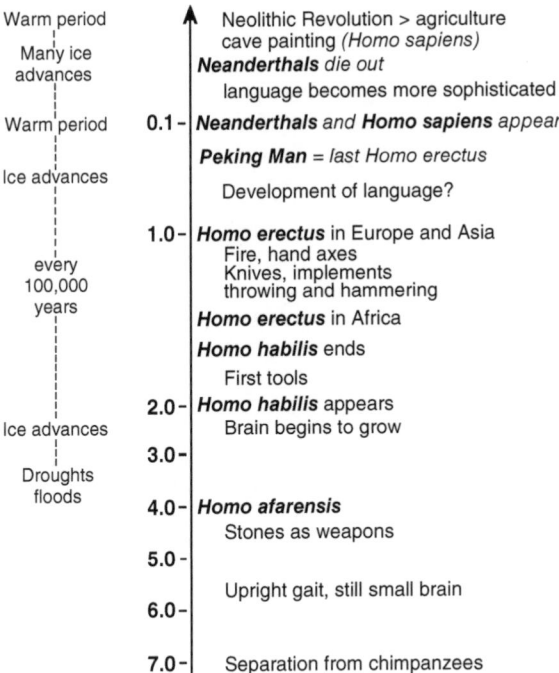

1.9 Hominid evolution, cultural history and climate.

cunning their disadvantages vis-à-vis more powerful carrion-eaters. For example, they used sharp-edged obsidian stones, which are easy to find in the volcanic areas of East Africa, to tear the flesh they needed. An upright gait helped them to detect and transport their carrion prey, so that a long and wide field of vision became a feature of the human species, in contrast to the keen sense of smell and hearing of true hunters. Other characteristics that developed in early humans were also part of this way of life: competition with more powerful carrion-eaters required stamina for running over long distances. In the climatic conditions of East Africa this led to loss of body hair. Nakedness and a special evaporation system based on perspiratory glands made it possible, in high temperatures, to maintain heat balance during periods of physical exertion.[70]

The development of the 'Homo' species – that is, of early humans – is thought to have been causally related to the Ice Age. The destruction of existing habitats was a threat in the Darwinian sense of the

27

term: it triggered waves of extinction in the plant and animal world and led through selective pressure to the development of new breeds.[71] Because of water binding in ice the climate was globally drier, even in the East African highlands, the cradle of humanity. Hominid evolution now divided into two lines: the especially strong australopithecenes, who could use the abundance of tough vegetable food; and the more delicate 'Homo'.

Meat consumption gave this second genus an additional evolutionary advantage, as brain growth assisted the further development of survival techniques. Homo habilis, a primate weighing some forty kilos, erect in much the same way as today's humans, had a brain of 500 to 800 cubic centimetres and already used simple stone tools (hence the name, habilis or skilful), with a capacity to fashion them through the technique of hard direct blows with a stone hammer. The number of items found suggests a conscious transmission of inherited knowledge, which represents for the first time an identifiable 'culture'. The use of fire, and therefore techniques of fire-making, are attributed to the Oldowan culture, named after the find in the Olduvai Gorge in Tanzania. Homo habilis, who carried tools over large distances, planned for the future.[72]

Climate change and the first globalization in the Palaeolithic Age

All these characteristics were even more pronounced in the successor, Homo erectus, who first appeared in East Africa about 1.8 million years ago with a brain size above 1,000 cubic centimetres, roughly 70 per cent of modern man's. Whereas the shape of their head seems primitive, the gait of these early humans was – as their name suggests – upright. Measuring 1.5 to 1.8 metres in height, they must have cut quite an imposing figure. Homo erectus, with an average weight of fifty kilos, was the first representative of the human genus to spread from Africa over large parts of the world, reaching east Asia more than a million years ago and leaving traces there as 'Peking Man' or 'Java Man' (find: 1.4 million years). The recently discovered Homo floresiensis probably also belonged to the family of erectines. The bone finds in Java, Flores, Borneo and the Philippines do not, however, prove the invention of travel by water, since these later islands were then, like Japan, still connected to the continent. Sea crossings to Madagascar, New Guinea and Australia could not yet be undertaken. The stretches that could be travelled on foot – from Africa to South-East Asia – seem immense, but at the rate of 20 kilometres per gen-

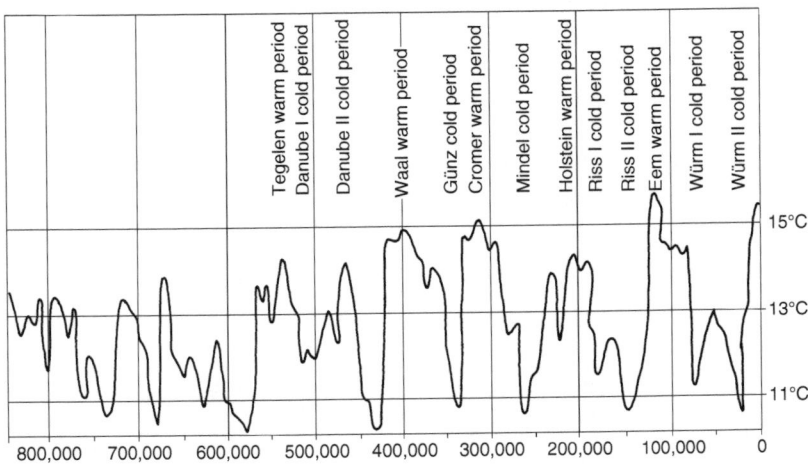

1.10 Temperature changes over the past 850,000 years. Sedimentation analysis from ocean drilling confirms the ice core results.

eration the distance from East Africa to East Asia could have been covered in 20,000 years.

A distinctive 'culture' has been attributed to this human group too. The so-called *Acheulean* culture (named after the archaeological site at Acheul, a suburb of Amiens, France) spread all over Africa and parts of Europe and West Asia (including India). Its 'techno-complex' features carefully worked hand-axes – a universal ice age device for hitting, digging and scraping – and cutting tools, as well as specialized wooden instruments and firemaking techniques. This evidence of intensive activity is why the term *Homo ergaster* ('labouring man') has been used for this early human type, which many palaeoanthropologists see as the true precursor of modern man. In Africa the Acheulean culture followed the Oldowan and took root some 1.5 million years ago. For Europe the finds begin approximately one million years ago and last until roughly 100,000 years ago. Interestingly, the culture of the erectines divided with the development of hand-axes, since the whole of eastern Asia and parts of eastern Europe stuck to the use of simpler stone implements. The world thus split into two technologically different cultures. The reasons for this split are not clear. It is possible that the Acheulean culture developed only after the emigration of Asiatic erectines.[73]

The spread of early humans across the planet accompanied a climate change that brought glaciation in the northern hemisphere

29

but increased rainfall in Africa. During this *pluvial*, forest advanced into savannah regions and reduced the living space for *Homo erectus*. At the same time, however, tropical Africa was no longer divided from the rest of the world by a wide band of desert. The Sahel and the Sahara became fertile regions that attracted further immigrants. This happened several times in the history of humanity, but the first time it was *Homo erectus* who benefited from the shifting zones of vegetation.[74]

Dramatic climate change required a great capacity for adjustment. People could respond either through migration or through local adaptation to the repeated mutation of rainforest into savannah, wet into arid conditions, heat into cold. The dispersal of familiar game favoured migration among hunters – or, at least, game animals appeared in Europe together with the first erectines. Migration made it necessary to cope with different climate zones. In addition, people had to become capable of alternating between animal and vegetable food, to become 'omnivorous'. They had to learn to detach experiences from particular places in their familiar environment and to find new, comparable locations. This power of abstraction placed greater demands on communication and led to development of the capacity for thought.

— 2 —

GLOBAL WARMING: THE HOLOCENE

Children of the Ice Age

The development of *Homo sapiens sapiens* took place during a phase of climatic history upon which we can today look only with horror. To put it in a nutshell, we modern men and women are 'children of the ice age'.[1] We speak of 'modern man' because, in the 1980s, it became possible to demonstrate that the whole of present-day humanity stems from a mother who must have lived 150,000 to 200,000 years ago in East Africa, the so-called 'mitochondrial Eve'. The proof came from the so-called mitochondrial DNA (mtDNA), which can be passed on only through the female line, and which is known for its regular mutations that allow us to determine the age of a subject. From the fact that the genetic substance of all humans alive today varies in Africa but is similar outside Africa, it was concluded that *Homo sapiens sapiens* could not have crossbred with older hominids such as *Homo erectus* or *Neanderthal man*. Otherwise, these would necessarily have left traces somewhere in the genetic make-up.

The said 'Eve', however, was not a single mother; we are probably talking of a population of roughly ten thousand who acquired selective advantages in the East African Rift Valley. This human type was stronger than earlier hominids – with the possible exception of the Neanderthals, earlier migrants from Africa, who also belonged to the *Homo sapiens* genus and who, as powerful hunters of big game, had perfectly adapted in Europe to the conditions of life in the ice age.

For some years now it has been supposed that a kind of primeval catastrophe, which largely wiped out earlier species of humans, is the reason why everyone living today stems from one original mother.

This would have reduced *Homo sapiens sapiens* to a few thousand individuals, with a corresponding impoverishment of the genetic pool to a single line. Geologists such as Michael R. Rampino (New York University) and anthropologists such as Stanley H. Ambrose (University of Illinois) blame this on a volcanic super-eruption: the explosion of Toba in the north of Sumatra island (Indonesia) approximately 75,000 years ago, which sent such quantities of ash and aerosol into the stratosphere that the clouds remained there for a number of years. The eruption is detectable today in ice core and ground sediment around the world.[2]

The entry of ash and aerosol into the stratosphere led to rapid cooling by as much as 15 degrees Celsius in particular regions and 5 degrees worldwide, for a period that lasted a number of years. The 'volcanic winter' interfered with plant growth and therefore had a negative effect on the food chain on both land and sea. Tropical vegetation must have been largely destroyed, and the forest seriously harmed in temperate regions. The surviving vegetation would have suffered a severe disturbance and taken decades to recover. The eruption of Toba had far more dramatic consequences than those of any more recent volcanic explosion.[3] This could explain, then, why the early history of *Homo sapiens sapiens* witnessed a diminution of the species that brought it to the verge of extinction.

Of course, with the recovery of plant life, the surviving humans had a unique potential for development, since the environment could be resettled and there were no longer rivals to limit their spread. There was rapid population growth, probably brought about by better adaptation of particular groups to their surroundings.[4] The further effects of the volcanic winter are unclear. Some have wondered, for example, how far the Toba explosion was responsible for the ensuing migrations among early humans and for a number of technical innovations.[5] Climatic feedback effects are another area for thought, since temperatures reached a low point in the following thousand years.[6]

'Beringia' and the globalization of humanity

Like earlier humans, *Homo sapiens sapiens* left Africa as a result of climate change. The forests pushed forward during the interglacial periods, and this caused large animals to withdraw from the savannah and gave a major boost to tropical diseases such as sleeping sickness and malaria. At the same time, during the pluvial periods, new forms of life appeared even in former desert regions, where the savannah that had retreated from tropical latitudes now re-established itself.[7] The forerunners of modern humans, like the Neanderthals before them, found a bridge across Palestine to Eurasia, and from there – like *Homo erectus* – to southeastern Asia. About 70,000 years ago, *Homo sapiens sapiens* spread through the whole of southern Asia.

One especially interesting chapter is the settlement of Australia. After *Homo sapiens* had crossed land bridges to the Indonesian archipelago and the dryland Sunda Shelf, a little more than 35,000 years ago, his successors reached 'Greater Australia', a continent encompassing today's Australia, Tasmania, New Guinea and the intervening Sahul Shelf. The width of the straits to be navigated was, in favourable conditions, approximately ninety kilometres. Already 35,000 years ago, Stone Age humans probably managed to travel across by raft, dugout or outrigger canoe. The calm sea and warm climate would have made it easier to survive for several days with an adequate supply of drinking water. The Australian aborigines are descended from these immigrants. By 30,000 years ago, flint was being mined at the southern tip of Australia, and Tasmania had been settled.[8]

There are various theories for the settlement of the Americas, but all have grown up around *Beringia*, named after the Danish navigator Vitus Bering (1681–1741), who in the 1720s was commissioned by Tsar Peter the Great to explore the eastern frontier of Siberia. *Beringia* is the term for the dryland bridge that appeared several times in the last ice age between Asia and America, and which is nowadays covered by the Bering Strait. At its narrowest point, between the Siberian Chuckchi peninsula and Alaska, this strait is only ninety kilometres across; but the climate makes it very unlikely that people could have crossed it on Stone Age rafts. The most favourable conditions for migration from Asia over the Beringia land bridge prevailed during several long glacial periods, when the sea level had fallen sufficiently low: that is, about 50,000 years ago, and again between 25,000 and 14,000 years ago. Life on the land bridge was not different from life in neighbouring Siberia. The climate was extremely cold and very dry – hence Beringia was ice-free. Pollen analysis has shown that the tundra-like summer vegetation of grasses, dwarf bushes and deciduous trees would have been sufficient to sustain a wealth of ice age fauna. Fossil finds indicate the presence of woolly mammoths, musk oxen, bison and caribou.

Asia's ice age hunters needed only to follow the big game to reach Alaska, and then, travelling across ice-free zones, to strike far south across the Canadian Shield. Opinions are quite diverse as to the exact period in question. Findings of human settlement 30,000 years ago have repeatedly been claimed in southern Chile or Brazil. That is conceivable in principle, but it would be very surprising in view of the fact that – as far as we know – the tundra regions of Siberia were first settled by big game hunters some 20,000 years ago, and that Beringia could have been crossed only from its easternmost point. Excavations in Dyukhtai, east of the Yenisey river, have dated the earliest cultural group there at 18,000 to 12,000 years ago. Evidence of this group reaches to the tip of the Chuckchi peninsula, where the ancestors of the first Americans lived – not counting dwellings on Beringia now lying under water.

The oldest verified archaeological finding of human settlement in North America, a mammoth bone from a cave in Canada from which the flesh has been stripped away, dates back 15,000 to 16,000 years. Skilfully worked microblades and other Stone Age implements, similar to those already known from eastern Siberia, have turned up from the period 15,000 to 12,000 years ago. Lush flora was present there at the end of the Ice Age, before Beringia disappeared beneath the waves. Probably there were already birch forests, and it may have been then that large groups migrated to North America. Nomadic hunters would have spread out from there, and it seems that already 12,000 years ago permanent settlers had taken possession of the cool southern regions of South America. With the flooding of Beringia, the early Holocene thus came to witness the first genuinely American cultural group: the hunter society of the *Clovis culture*, so called after an archaeological site in New Mexico in 1932. These mammoth-hunters were ancestors of the people that Europeans called Indians thousands of years later. The globalization of humanity was facilitated by the deep sea levels of the last great ice age. After it ended, the Eurasian, American and Australian cultures developed further in isolation from one another.[9]

Europe in the Würm glaciation

It would seem that some 50,000 years ago, having learned to adapt to colder conditions, *Homo sapiens sapiens* spread north from Palestine towards Asia and, some 40,000 years ago, to ice age Europe – which was then possible across a wide land bridge at the Bosphorus. Asia Minor was tightly attached to Europe: the Black Sea was an inland sea. The modern human type, with high forehead, small teeth and receded eyeballs, has also been known as *Cro-Magnon Man*, after a find in the Dordogne, at the Abri de Cro-Magnon. The skeleton is identical with that of contemporary humans.

The climate was then quite harsh in large parts of Europe. The north was covered with a thick armour of ice, which stretched from the North Pole to the north German lowlands. Glacier formation had so lowered the sea level that the British Isles were part of the continent. All over the world glaciers thrust down from the mountains into low-lying areas, carving valleys ('glacial valleys') and forming basins in which the big lakes we know today took shape at the end of the last ice age. At the high point of the last ice age, approximately 20,000 years ago, the major rivers ran dry. Glaciers bound the water, and the ground was frozen to great depths. All this sounds horrific. But the latest research draws a surprisingly friendly picture: living conditions in Europe were even especially favourable for man. The climate was distinguished by its great stability. The weather was far less changeable than it is today. The summers were mostly mild and pleasant. In winter it remained frosty throughout, but there were no Arctic temperatures and the weather was generally dry. Ice age winters may be compared to sunny January days in the Alps, when it is possible to sunbathe even in the depths of winter. Average temperatures were four to six degrees lower than today, but the dry air meant that that they were not unpleasant. Spring came late, but summer temperatures climbed to around 20 degrees Celsius.

The low temperatures kept the vegetation severely limited, but the ice age tundra of Central Europe bore no resemblance to tundra inside the Arctic Circle. At this latitude the sun's rays remained strong, the summers were warm and lush vegetation, well supplied with meltwater, offered ample food to animals. The tundra did not lag behind the East African savannah in the wealth of its big game. It permitted the development of a megafauna, which included not only plant-eating mammals such as the mammoth and woolly rhinoceros but also the aurochs or urus, elks, deer, cave bears and predators such as lions and hyenas. Early man was able to live on their carrion meat, which could easily be frozen and stored in 'natural refrigerators'. But a new capacity was also developing – for big game hunting.[10]

The first Europeans and the birth of art

In the Upper Palaeolithic, the latest part of the Old Stone Age, the cultivation of distinctive productive styles began to divide human history into style periods and geographically demarcated 'cultures'. Thus, for the period around 40,000–30,000 BC, we find in the whole of inhabited Europe a unified culture that has been named *Aurignacian* (after an archaeological site in France). This culture displays the earliest blade technology, in which bone points have grooves cut in the bottom to help keep them fast. The knives and blades are sharpened. This somewhat warmer phase of the glaciation also saw the earliest art, or, to put it more emotively, the birth of art.

Alongside its carefully worked stone implements, this culture surprises us with its multiplicity of fine bone statuettes of objects such as cave bears, probably produced while they were hibernating. Ornaments have been found in great quantities: pierced animal teeth, stone and ivory beads, and drilled snail or mussel shells partly imported from the Mediterranean. A characteristic piece of sculpture is the famous Venus of Willendorf (from the Wachau valley in Austria), which places greater emphasis on the woman's fertility attributes than on a representation of her head. As far as Germany is concerned, many discoveries have been made in Baden-Württemberg, including the Geissenklösterle cave in the Swabian Alb, with its ivory mammoths, cave bears, bison, horses and one standing human figure. The Hohlenstein Stadel cave in the Lone valley gave up one of the most important works of ice age art: a standing man with leonine head.[11] This statuette permits inferences about religious conceptions at the time, as it refers to the idea of animal metamorphosis and hence to the shamanist belief system.[12]

From this oldest phase of art come the flamboyant cave paintings discovered a few years ago in the Dordogne, in the Grotte Chauvet. In this 'Sistine chapel' of the Ice Age, the whole range of large fauna is placed in relation to man as hunter. The complex artworks, in inaccessible caves that had to be laboriously illuminated with torches, point to new ways of conceiving space, since the lavish places of worship could scarcely have been designed for the fleeting moment. Probably Stone Age hunters kept returning there for acts of worship or initiation rites. The previous occupants had been hunter nomads, who followed the migration routes of big game. But the cave painters now firmly associated their cult with fixed locations. They were the first Europeans.[13]

Cro-Magnon Man survived the glacial maximum of 20,000 years ago in the ice-free southern regions of Europe. The Solutrean culture (named after the site at La Solutré in France, approx. 22,000–18,000 BC) was well adapted to this especially harsh period. Its distinguishing features include the heat treatment of rocks, leaf-shaped spearheads and a pressure flaking technique for stone tools. The find of bone pins demonstrates that they were capable of producing suitable

36

clothing and tents out of animal skin. Finds at Dolní Věstonice in today's Czech Republic show that dwellings were dug as far as one metre below ground, so as to seal the walls better against winter storms. The walls were made of wooden stakes covered with animal skins. Mammoth bones were used along with wood for heating purposes. In Ukraine settlements have been found where mammoth bones were used in the construction of longhouses twelve metres in length. Population density was generally very low. For the period of the glacial maximum, the total number of people living in France – the most densely populated region of Europe – has been estimated at no more than 2,000–3,000.[14]

Between Ice Age and Holocene: the culture of the Magdalenians

After the end of the last glacial maximum, the climate began to change everywhere in the world. It became warmer and drier, partly through those temperature leaps that are known as Dansgaard-Oeschger events (after their discoverers, Willi Dansgaard and Hans Oeschger). Especially in Europe, but also in northern Asia, the flora and fauna spread northward as soon as the glaciers retreated. This opened up new living space, not only in a geographical sense but also in terms of the seasons; the annual period of vegetation grew longer in already inhabited regions. This led to the development of a new culture, which overshadowed all earlier styles in wealth, even if continuities are also clearly visible. The *Magdalenian culture* stretched from 18,000 BC to 10,000 BC, from northern Spain across the Dordogne (where the eponymous site of La Madeleine is located) and on to Central Europe and Russia. More than eighty per cent of all known cave paintings come from the period between 15,000 and 12,000 BC – for example, in the caves of Lascaux, Peche Merle (Dordogne) and Altamira (Spain). The Magdalenian hunters were semi-nomadic and possibly began the process of domesticating animals.

Population density was still very low, although in France numbers increased threefold after the glacial maximum to reach a total of 6,000–9,000. One imagines that, like modern nomads, they lived in clans of 20 to 70 people, because this allowed conflicts to be kept within bounds. These groups were probably further organized into larger associations or tribes of 500 to 800 people each. According to the evidence of skeletal remains, average life expectancy was less than twenty years. Only 12 per cent (and no women at all) lived beyond the age of forty. The skeletons are characterized by defects and signs

37

of physical injury. Presumably there were already social hierarchies or strata, since graves in Russia and Italy have given up thousands of ivory and tooth beads that must have adorned the clothing of those laid to rest. In the case of children, in particular, this cannot be attributed to any merits of their own, but must have been due to their line of descent. It was still during the Palaeolithic that the previous egalitarian societies came to an end and status symbols began to play an increasing role.[15]

The end of the megafauna

This culture ended with the beginning of the Holocene, when the existential basis of the ice age hunters disappeared with the extinction of large fauna. The causes of this species extinction of large mammals are hotly debated in the literature. In one view, ice age hunters waged a kind of blitzkrieg against them, pursuing them to the far corners of the earth and slaughtering them in a global 'prehistoric overkill'.[16] Against this, however, is the fact that the megafauna did not die out everywhere. Elephants survived in Africa, India and South-East Asia, as did giraffes, rhinoceroses and wild cattle. Wild horses did become extinct in America, but they survived in Asia – along with cattle – until they became domesticated by humans.

The extinction of the woolly rhino, the great cave bear, the European sabre-toothed tiger and the mammoth is today overwhelmingly regarded as a consequence of climate change. We know from pollen charts of high-moor bogs that the forests advanced at the end of the Ice Age. The habitat of the megafauna – ice age tundra in the southern latitudes of Eurasia and North America – disappeared, and mammoths and woolly rhinos, giant deer and wild horses lost their staple source of food. The tundra withdrew into Arctic regions, and the large mammals had to follow suit. But the climatic conditions in the Arctic were much more severe, because winter temperatures there were considerably lower.

Further south, it was not so much rising temperatures as increased aridity that spelled trouble for large animals. We know from remains in the Siberian permafrost what was the selective disadvantage of mammoths: their hide was sensitive to moisture, since, like today's elephants, they had no sebaceous glands to lubricate their hair. This was of no consequence during the Ice Age, but now it became a real problem. Their hair absorbed moisture and dried out only with difficulty. A wet mammoth sinks into the bog – precisely where frozen remains have been discovered. Similarly, in a climate that is growing

38

warmer, musk oxen are susceptible to heavy colds and fatal lung infections. It has been estimated that the population of large animals declined by 99 per cent as a result of climate change. What is unclear is how mammals had been able to survive the previous interglacial periods; the *Eem interglacial*, for example, had been scarcely less warm than the Holocene. Perhaps, after all, the appearance of new humans and their hunting methods did play a role in the mass extinction.[17]

Global Warming and Civilization

'Man appears in the Holocene' is how the Swiss writer Max Frisch (1911–91) put it.[18] But, as we have seen, humans were actually 'children of the ice'. We might say rather, with Stephen H. Schneider and Randi Londer, that the global warming of the Holocene brought about 'climates of civilization'.[19] For, even if it may sound strange in the context of today's discussions, it was the global warming of the Holocene that smoothed the way for advanced civilizations. The concept of the Holocene was first introduced in 1885, at the International Congress of Geologists, to designate what geologists saw as the 'quite recent period' of the past ten thousand years, which has been marked by a warmer ice age climate.[20] In the perspective of cultural history the Holocene really does constitute a unity, for its novel forms of human culture developed into the civilization that we know today from our own experience.

In the Holocene *Homo sapiens sapiens* began to make massive incursions into nature, turning it into a cultural landscape. At the beginning of this period agriculture and livestock breeding developed in a number of favoured regions; nomadic hunters built up fixed settlements. The 'Neolithic Revolution' – the upheaval in ways of life at the beginning of the New Stone Age (Neolithic) – brought the first deliberate production of foodstuffs, with improved techniques of food preparation, storage and house construction. Societies became more differentiated and stratified as the first cities emerged as the kernel of the so-called ancient civilizations. The population of the world increased, from an estimated base of five million at the beginning of the Holocene.[21]

The first temples in the golden age of the Allerød

In the transition from the last great ice age to the Holocene post-glacial, the climate was persistently cold and dry, interrupted by warm intervals. In the Allerød period (named after a Danish archaeological site), which began about 12,000 years ago (= 10,000 BC), the forests began a new advance under the impact of significant warming and increased humidity. The Magdalenian culture spread northward and became more diverse. An especially important find was a base camp with round huts averaging 6 to 8 metres in diameter, which were heated by open fires on stone slabs. Upright posts supported conical roofs, which, like the 'walls', were probably made out of horse coat. Cooking was done in special pits with the help of heated slabs of stone. The houses were inhabited only seasonally, as the hunters had to follow the animals where they went. In common with earlier Old Stone Age cultures, the Magdalenian was still dominated by the culture of big game hunters. However, horse and reindeer were already the most common bag. Artistic production included ornaments, statuettes of animals and women, and geometric symbols. This culture ended with the beginning of the Holocene, when its existential basis disappeared with the extinction of large fauna.[22]

Meanwhile, a complete transformation of human ways of life was emerging in the Middle East. Here too, before becoming sedentary, *Homo sapiens* had established places of worship to which Stone Age hunters and gatherers regularly returned. Recent excavations have yielded particularly amazing discoveries in the Middle East, the future cradle of advanced civilizations. The monumental *Göbekli Tepe* ('navel mountain') site in Anatolia, the oldest temple complex in the world, has been dated by its German excavator, Klaus Schmidt, to a time 12,000 years ago.[23] Fixed places of worship had already existed for millennia in the painted caves, but the construction of stone monuments with Stone Age tools, in a geometrical, circular layout, involved an entirely new kind of social achievement. This can probably be related to a more complex form of social organization, and perhaps also to changes in the realm of religion. It may be that these are even linked to positive climate change during this period: people must have been grateful to the heavenly gods for the sudden improvement in soil fertility. And they could have best expressed this in a place of worship open to the heavens, on a mountain to which people streamed from far and wide.

Fixed settlements are found a little later. The US archaeologist Stephen Mithen has shown, in his survey of postglacial excavations, that the whole *Natufian culture* – so called after Wadi Natuf in Palestine – consisted of Stone Age settlements that did not yet have any agriculture, livestock breeding or pottery. Sickles with obsidian blades were found in the villages, but there was no evidence of planned cultivation of cereals. The skeletons and teeth of the settlers showed no signs of inadequate or defective nourishment or of injuries from fighting. This allows us to conclude that their living conditions – that is, the climate and environment – were so favourable that they could use fixed settlements as a base to hunt game and harvest wild cereals. Until recently that would have been considered an impossibility. But the model was evidently so successful that it soon led to a population increase and the spread of Natufian villages through the regions of present-day Israel, Syria, Iraq and southern Turkey.[24]

Renewed cold and cultural relapse in the Younger Dryas period

This paradise on earth came to a rather abrupt end some 11,000 years ago (= 9000 BC), when the climate grew colder and drier. The temperature in central Greenland fell by 15 degrees Celsius, and in the region of today's Poland still by 6 to 7 degrees.[25] In central Europe the sub-Arctic climate returned for more than a thousand years – the last time this has happened. At the same time, the *Dryas octopetala* flowering plant advanced once more in the north of Germany. The flora brought back a matching fauna and the human culture that corresponded to it: only a hunter culture could exist on that meagre foundation, and reindeer provided the crucial source of food. In cultural terms, this was the last phase of the Old Stone Age, the Upper Palaeolithic.[26]

The cooling had effects as far away as the Mediterranean. In the Middle East all the Natufian settlements had to be abandoned. With the return to nomadic hunting, population figures must have fallen dramatically. The climate had an impact here on the development of culture and perhaps even on much of its substance: it was no accident that punitive weather gods later played a crucial role in the pantheon of the Middle East. This first becomes visible with the development of writing, but its roots probably go further back, perhaps to the loss of paradise in the Younger Dryas Period. In any event, the Sumerian storm god Ishkur/Adad is already in top place in the oldest surviving list of deities.[27]

Holocene global warming and changes in nature

After a thousand years, the Younger Dryas ended as abruptly as it had begun. Within a few decades the onset of the Holocene raised

41

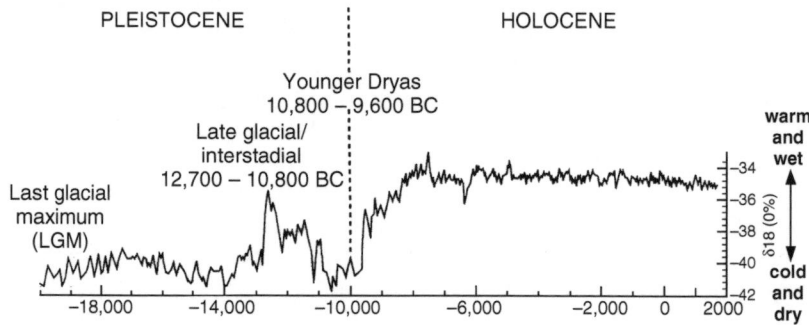

2.1 Global warming in the Holocene. The proportion of oxygen isotopes reveals dramatic temperature leaps, until the climate stabilized 10,000 years ago.

average annual temperatures by as much as 7 degrees Celsius. Storms grew less intense and rainfall doubled. It is, as always, ultimately unclear what led to this radical climate change, this global warming. In discussions, increased solar activity is mentioned as the primary trigger.[28] As soon as the warming process kicked in, all possible feedback effects began to happen – from a diminution in albedo to a changed composition of the atmosphere as vegetation spread in northern and southern latitudes. Only during the Holocene did the environment we now think of as 'natural' make its first appearance. Through the rise in sea levels, the continents acquired more or less the shape they still have today, and flora and fauna alike adapted to the new climatic conditions.

Biblical deluges and changing coastlines

As the glaciers melted, the coastlines began to change. One fine day about 8,400 years ago, an observer on the Bosphorus would have heard a gurgling noise that ushered in one of the greatest flood disasters in human history. During the great ice age, the level of the sea fell so low that Europe was connected to Asia by the Bosphorus. The Black Sea became a great freshwater lake, fed by the Danube, the Dneiper and the Don. Hunter-gatherer cultures had been present for millennia on its flat shores. Fisherfolk lived in Late Stone Age villages; farmers had cleared trees, dug arable land, built homes for themselves and enclosures for their livestock. But the coastline of the isolated Black Sea lay more than one hundred metres below that of the Mediterranean, and the melting ice raised its water level more quickly. This put ever greater pressure on the blocked straits and at some point broke through the Bosphorus. Salt water began to rush down into the Black Sea with the force of several hundred Niagaras; the roar must have been audible at a distance of hundreds of kilometres. Salt water flowed into the basin for months, eventually filling it to the level of the sea. Hundreds of square kilometres of inhabited land were lost in the catastrophe. The remains of the early Black Sea cultures now lay buried underwater.[29]

Rising sea levels changed the shape of coasts everywhere. The Beringia subcontinent disappeared, and so did the land bridges that had joined mainland Asia with Japan and with the large Indonesian islands (the Sunda land bridge). The links between Australia and New Guinea, India and Ceylon, Africa and Madagascar sank for ever beneath the waves. Large coastal regions that had previously been favourites for hunting and settlement vanished in most parts of the world. New straits suddenly appeared (for example, the Bering Strait, the Sunda Strait and the Sea of Marmara), as well as new inlets such as the Persian Gulf or the Red Sea. New seas such as the Baltic or Hudson's Bay arose in place of earlier glaciers. Approximately 9,500 years ago (7500 BC), the English Channel was formed and cut off Britain and Ireland from the continent. Roughly 8,000 years ago, in a violent natural catastrophe, the sea broke into Hudson's Bay. And some 7,000 years ago the flooding of Dogger Bank began in the North Sea. Sicily was separated from Italy, and the Greek islands from the Anatolian mainland. The sea put an end to the coastal cultures and drove men and women inland.

Transition to the Mesolithic

The global warming is associated with a fundamental shift to a culture with more diverse and sophisticated features than before: the transition from the Old Stone Age (Palaeolithic) to the Middle Stone Age (Mesolithic), the last hunter-gatherer culture in the lands of Europe and the later advanced civilizations.[30] The significance of the warming at the beginning of the Holocene has always been beyond question. Already in the 1960s we could read in the Propyläen world history edited by Alfred Heuss and Golo Mann: 'The transformation of the Upper Palaeolithic into the Mesolithic form of society took place relatively quickly; it was caused by climatic upheavals.'[31]

Global warming spelled the end of the previous form of economy. Sedentariness developed as the megafauna died out, since forest animals – after their initial migration – remained tied to a particular place. To hunt them required new techniques: the widely produced *microliths* (finely worked stone implements) are characteristic of the Mesolithic. Food traditions could be most easily preserved in the proximity of lakes and rivers. Probably that is why a majority of Mesolithic camps and dwellings lie close to water; it could also provide them with drinking water and guarantee a minimum of hygiene and waste disposal. Vast quantities of discarded mollusc shells testify to the constant presence of human beings. They were hunters and gatherers, but fruits such as berries and nutritious hazel nuts now formed a much greater part of their diet. The Munich palaeobiologist Hansjörg Küster even speculates that hazel bushes may have been deliberately protected and planted – which would explain their sudden spread around 9,000 years ago (= 7000 BC). In this view, humans consciously intervened in the growth of vegetation and began the process of transforming nature into a cultural landscape. We know little about the demographic consequences of this. It may be that the population on the fringes of postglacial forest was smaller than that which had lived on ice age tundra. The limits to growth were anyway narrowly set.[32]

The climate optimum of the 'Atlantic period' and the 'Neolithic Revolution'

It became wetter in the Middle Holocene, some 8,000 years ago. This 'middle warm period' is generally known as the 'Atlantic period' (approx. 6000–3000 BC).[33] It plays a special role in our history because – long before man could exert a greater influence on nature – it was by far the warmest and at the same time longest phase of the Holocene. Temperatures were on average 2 to 3 degrees higher than in the late twentieth century. Glaciers melted on a wide scale and released large quantities of water. A more humid climate prevailed in the whole Middle East, right up to India and China. The levels of seas and lakes around the world were higher than they are today. Lakes such as Lake Chad in Africa took on the dimensions of inland seas, and the Nile floods were as much as seven metres higher than at the time when the Aswan dam was built. These changes led to a blossoming of North Africa. In the central Sahara, with its frequent rainfall and plethora of lakes and rivers, the dense early Holocene population of big-game hunters was gradually replaced by nomadic herdsmen. It may be that cattle were domesticated here for the first time in history.[34]

As the term 'climate optimum' suggests, this warm humid 'Atlantic' was especially favourable for the development of human culture. There were crucial improvements in technical equipment, which was still mainly produced out of stone, although other materials were also increasingly employed. The stone axe replaced the Palaeolithic hand axe as the most important tool.[35] This was the phase of transition to the 'New Stone Age'. The *Neolithic* represents a decisive period in the history of humanity: the passage from the semi-nomadic hunter-gatherer culture of the Middle Stone Age to a sedentary culture of farmers and livestock breeders. Greater simplicity of food provision may have been the first inducement to change. But the growing population made it necessary to cultivate the land in a purposeful manner, and this once again – only now lastingly – expanded the ecological habitat available to humans. The transition to agriculture occurred in the Middle East 10,000 to 9,000 years ago (= 8000–7000 BC) and in parts of Europe approximately 8,000 years ago.

2.2 Sediment remains in the central Sahara testify to the presence of large lakes during the early Holocene. In warm periods, there was more water in the planetary water cycle, and the monsoon covered new areas.

This brings us to the origins of human civilization: the Latin noun *cultura* (which, like one sense of the German *Kultur*, is synonymous with 'civilization') is in fact derived from the verb *colere, cultum*: to cultivate, to inhabit, to worship. The passage from hunting to agriculture was of such fundamental importance that some have compared it to the Industrial Revolution. Using this analogy, the Australian archaeologist Gordon Childe (1892–1957) coined the term 'Neolithic Revolution' in 1936. And, even if transitions are today viewed more fluidly,[36] it can certainly be argued that no similar development took place in the conditions of the Ice Age.

The Neolithic Revolution liberated humans from the great insecurity associated with hunting, fishing and fruit-gathering. The purposeful cultivation of plants, which led through seed selection to the growing of new strains, brought about a fundamental change in lifestyle such as one never finds in the biological evolution of species. The capacity for such change is actually distinctive of the human species. Close ties to the soil, which have to be maintained at least from sowing to harvest time, set up a strong tendency to sedentariness. This helped to optimize agriculture and to promote the domestication of wild animals, beginning with sheep and goats, for farm or household use. We know today from DNA analysis exactly where the domestication of animals and the transition from wild cereal-gathering to purposeful agriculture happened. As Stephen Mithen has shown in his magnum opus, *After the Ice*, the genetic make-up of modern wheat most strongly resembles that of wild strains in an area of southeast Turkey less than thirty kilometres from Göbekli Tepe.[37]

It would seem to be here that the Neolithic Revolution began. This cradle of human civilization, whose significance was first recognized only in the 1990s, lies just a little north of the region traditionally known as the Fertile Crescent. For man's intensive engagement with nature took place not in the lowlands but in the high hills further north, in the foothills of the Taurus and Zagros mountains. Livestock breeding also probably originated here. The animals most useful to man – sheep and goat, pig and cattle – first became accustomed to human handling, and were then continually used and bred, some 9,000 years ago (= 7000 BC) in western Asia. The southwestern region of present-day Turkey, together with Iraq, Syria and Israel, is where civilization had its origins. Agriculture and livestock breeding jointly expanded the food margin and placed human existence on a more secure footing, especially when – in the sixth millennium BC, at the latest – animals were also deployed for the tilling of the soil.[38]

Rice cultivation and the Chinese landscape

With the beginnings of agriculture, Neolithic man intervened in the natural environment. Palaeolithic hunters already probably used fire for hunting purposes and brought about extensive changes in the landscape, both in Eurasia and in Australia or North America. Clearance by fire released large quantities of carbon dioxide – although it is impossible today to determine the ratio of 'natural' forest and bush fires to those deliberately started by humans. Neolithic encroachments in the landscape reached a new dimension as fresh areas were permanently cleared for settlement and for arable land or pasture. The advance of agriculture was not without consequences for the character of the vegetation. Pollen analysis in the British Isles, for example, has shown that the forest clearance was already concluded in Neolithic times in the southern lowlands and Ireland, but that hilly regions and the Scottish Highlands were first cultivated only in Roman times or the Middle Ages.[39]

In western Asia, Europe, northern India and the Indus Valley, agriculture meant first and foremost the growing of cereals, whose fruits could be processed with water into porridge and immediately eaten. Grain could be fermented to make beer or ground into flour and baked as bread. For many thousands of years bread would remain man's staple food, revered also in religious texts: 'Give us this day our daily bread.' The multiple origins of the Neolithic Revolution show, however, that wheat and barley were not grown everywhere as staple crops. Millet and sorghum took their place in Africa and southern India, or maize in the Americas. These cereals were each taken from the wild and purposefully sown in fields, initially without any major effects on the environment. In this respect, the cultivation of rice in paddy fields was fundamentally different.

Rice plants have to interact with water for many months, and the intricate systems of irrigation and drainage developed by farmers release gases (including methane and, of course, water vapour) that have a major impact on the climate. The digging of terraces for cultivation radically changed the landscape over wide areas. This form of rice-growing required considerable investment and had implications for the organization of society. The crop is thought to have originated in southern China, as far back as the beginning of the Holocene, according to recent research. Grains of rice are compact, nutritious and easy to store. It is therefore assumed that they were exported at an early date. Archaeological finds of rice grains in other

47

regions – in northern China, for instance – may point either to imports or to independent cultivation. Around 3000 BC they arrived in Thailand, Vietnam and Taiwan, in association with forest clearance and the cutting of terraces. Rice-growing reached the Ganges valley, Indonesia and Malaysia around 2500 BC, and Korea and Japan around 1000 BC.[40]

The systematic cultivation of rice permitted a demographic leap forward, with all its advantages and disadvantages. The most important advantage was the increased density of cultural traditions and the development of advanced civilizations. Since the Neolithic period southern China has been the most densely populated region in the world. Its civilization stretches back to 2800 BC – even if the early dynasties have a legendary side to them.

The stable warm period as the basis of the ancient civilizations

If the myths of a 'Golden Age' have a basis in reality, they may refer to the persistently warm climate of the New Stone Age and the Bronze Age. Climate historians consider that this period was largely free of stormy weather, and that the stable climate promoted a rapid and extensive development of trade and cultural intercourse. The land and sea routes of antiquity took shape during these times. New mountain passes and raw material deposits became accessible. Tin from England and amber from the Baltic were traded as far afield as the Mediterranean. The settlement of remote areas is demonstrated by the prehistoric megalithic culture, which stretched from Stonehenge through numerous sites in Ireland to the Hebrides and Orkneys. The climate then must have been considerably more favourable than today, as astronomical systems to observe the winter solstice make little sense unless it was possible to count on clear weather to study the heavens. There must have been less cloud cover than in the past millennium. Historians assume that there was a northward shift in the zone of high pressure during the Neolithic climate optimum.[41]

Some authors attach as much importance to the 'Urban Revolution' as to the Neolithic Revolution, on the grounds that it symbolizes the passage from agrarian society to advanced civilization. They portray a denser pattern of settlement within peasant society, one that rested upon a more extensive division of labour and freed ever larger numbers of people from basic production for other tasks in the economy, religious life, administration or defence: people such as priests, the king and his royal court, state officials and employees, craftsmen, merchants and soldiers. Greater specialization probably developed in basic production too (farmers, herdsmen, fishermen, etc.), and in addition there was still the division of labour between the sexes. In most cultures, women participated in field labour and often in market life. In any event, urban civilization was based upon the release of a large number of people from activity in agriculture.

Multilayer excavations in Jericho going back as far as 7000–8000 BC suggest that the first cities began as large villages and gradually developed urban structures. The prerequisite for greater settlement concentration was a rise in population numbers that could be sustained through a correspondingly productive farming economy. Urbanization led to an accumulation of central powers and in some cases to new legal forms that separated town from country. One visible sign of this separation was the wall (in Jericho approx. 7000 BC), which had legal as well as military significance and therefore stood even in modern times as the true symbol of the city. Urban life was dominated by men representing rulership, religious life and the specialized trades. Whereas social stratification is possible even in peasant societies, only urban civilization with its much greater social differentiation allows the exercise of power to be institutionalized. Advanced urban civilizations produce identity-building status symbols and, with the development of writing, a more or less permanent tradition. They are there at the beginning of history as we know it, in Egypt as well as in Mesopotamia, India, China, Mexico and Peru.[42]

Without wishing to revive old-style climatic determinism, one cannot but note that the ancient civilizations – the Mediterranean, Mesopotamia, northern India, northern China – lay at roughly the same latitudes, between 20 and 40 degrees north, quite far from the extreme climates of the tropics or the colder regions in the north and south of the planet. These milder latitudes offer a number of advantages: a reliable supply of water, sufficient warmth for cultivation of the soil, absence of great heat or long cold winters and deadly illnesses.[43] The core of civilization in ancient America was situated closer to the Equator, but also outside the tropics or in highland regions. It was based not on the cultivation of river valleys but on other forms of irrigation technology. The term technology already alludes here to the fact that knowledge of agriculture must have been highly developed. On closer examination, we can see that all the advanced civilizations presupposed a specific cultivated plant, which made possible the leap in population numbers on which urban civilization rested.[44]

The drying up of the Sahara and the rise of Egypt

The warmth and humidity of the postglacial optimum blessed not only northern latitudes (or southern latitudes in the southern hemisphere) but also arid regions of the planet. In so far as we can trust radioactive carbon dating, this situation lasted down to the time of the earliest Egyptian dynasties.[45] The drying up of the Sahara, which began about 5000 BC, points to an at least regional climatic change. Many experts in the field see a direct link between the reduction in the habitable area of North Africa and the quite sudden emergence of the first farming villages in the Nile valley, between 5000 and 4500 BC. The beginnings of agriculture in the Nile flood plain soon led to a rise in population. That underlies the history of the advanced civilization in ancient Egypt.

Originally the farming villages were small and simply structured. As they grew, little towns took shape and gave rise to capitals of small kingdoms that bitterly fought with one another for supremacy. Around 3200 BC one of the rulers in question succeeded in politically unifying the Nile valley north of Aswan. The civilization of ancient Egypt was born out of the subsequent fusion of cultural characteristics.[46] Egypt's Nagada II and Nagada III cultures, which spread from the south across the north of the country, represented a first period of florescence. Images on clay pottery from this time bear an astounding resemblance to the pictograms in rock drawings in the Sahara. With political unification came the development of a script, some two hundred years before the time traditionally attributed to the legendary First Dynasty. We reach historically more solid ground in the so-called Third Dynasty (c. 2640–2575 BC), which is usually seen as the start of the Old Kingdom (2640–2134 BC). The fact that the building of monumental pyramids immediately began with the second king, Pharaoh Djoser (reigned 2624–2605 BC), shows that this culture already had a long experience behind it.[47]

The foundation of Egypt's great power was its distinctive economy: the annual flooding of the Nile valley depended on summer rainfall in the Ethiopian highlands, which caused the river in Egypt to burst its banks in September and return to its bed only in October. Agricultural land received the necessary water, the residual silt from the Nile acted as a natural fertilizer, and as the water drained away it prevented the soil from becoming too salty. By the time of the First Dynasty the water level was being forecast by means of the so-called Nilometer, and its distribution in the fields regulated through basin dykes. The strength and cultural continuity of ancient Egypt should probably be attributed to the way in which the central kingdom exploited the regularity of the Nile's flow. This may already be seen in the demography. From the early dynastic period down to the Ptolemaic age, the controlled use of water made it possible to feed some 1.5 to 2 million people, a huge figure in comparison with all previous and most contemporaneous cultures. The high population gave the Old Kingdom pharaohs a base to fund major works and to expand their power to the south, west and east, into Nubia, Libya and Palestine.[48]

Ötzi and the later warm period

In the millennia of the New Stone Age, clearances and economic activity converted the forest of Central Europe into a 'man-made' landscape. The earliest settlements were already anything but primitive. The precisely crafted stone axes, with sharp stone blades, were well suited for working on wood. Progress was made in the domestication of animals, and the breeding of cattle, goats, sheep and pigs considerably expanded the room for manoeuvre in respect of food supply.[49] After excavations of quite uniformly designed settlements all over Europe, it has been possible to infer that the cleared areas were neatly divided between arable land and pasture, fenced off from farmers' homes and refuse ditches. As the population increased, more and more areas were cleared. What conservationists take to be 'nature' requiring protection has since the New Stone Age been the product of purposeful activity on the land – from regulated or unregulated river landscapes to high Alpine pasture.

During the Atlantic period the Alps were largely free of ice. Only as it came to an end did the mountain heights again gradually ice over. The find of the glacier mummy 'Ötzi' may serve here as a climate marker. This hunter from a valley in the South Tyrol, who was caught in a blizzard 5,300 years ago after he had crossed the Tisenjoch pass near the Similaun peak, was finally freed from the glacier in September 1991. The iceman's state of preservation indicates that the glacial snow did not melt during the medieval warm period. According to Konrad Spindler, the director of scientific research, it must be accepted that six autumn days in 1991 were the first chance in the last five thousand years for the glacier mummy to

51

be found. In that unusually mild year Tyrolean glaciers gave up another five persons – as many as in the whole of the previous forty years. But only Ötzi was still lying *in situ*, at the place of his death in a mountain hollow, undisturbed by the flow of the glacier.[50]

The collapse of the advanced civilizations around 2150 BC

The breakdown of the first advanced civilizations can be correlated even more directly than their rise with fluctuations in the climate. The crisis of the Old Kingdom in Egypt and the beginning of the 'First Intermediate Period', around the year 2150 BC, has been associated with failure of the Nile flood and therefore with one of the peaks of the Sub-Boreal. The climatic consequences did not determine the direction in which things developed, but 'they did rule out a continuation of previous subsistence paths'.[51] At the end of the long rule of Pharaoh Pepi II (2246–2152 BC), food emergencies broke out and led to a situation of dire poverty. The kingdom fell apart: the central authority no longer exercised any control. Pharaohs who were unable to ensure the fertility of the land lost their political legitimacy. Egypt was divided during the following dynasties, until it was again reunited as the Middle Kingdom, more than a hundred years later, under the 11th Dynasty.[52]

Things did not look much different in Mesopotamia. The rise of this civilization was itself due to climatic change. Following the Ice Age, a 110 metre rise in sea levels significantly altered the coastline of the Persian Gulf. At the end of the Atlantic period, around 3500 BC, the sea extended into the region of Ur, the centre of the Sumerian civilization. Ur and Eridu lay on a raised promontory. With the increasing aridity of the Sub-Boreal, the coastline receded and permitted human settlement on the fertile alluvial land.[53] One of the oldest epics in world literature describes the draining of the marshes. The king of Uruk speaks to the goddess Innana: 'There was a marsh then in Uruk. . . . Divine Enki who is king in Eridu tore up for me the old reeds, drained off the water completely. For fifty years I built, for fifty years I was successful.'[54] Like the rise of Mesopotamian civilization, its decline too is associated with extreme climatic events, the periods of drought at the height of the Boreal. The Akkadian kingdom, which politically unified the twin river country under King Sargon I (reigned c. 2371–2316 BC), collapsed at almost the same time as the Old Kingdom in Egypt (c. 2150 BC), to the accompaniment of rebellions by city-states and nomadic tribes.[55]

The kingdom of Akkad had a sophisticated system of water regulation and could draw on granaries to offset annual variations in rainfall and harvest yield. Nevertheless, northern Mesopotamia now had to be completely abandoned, and the south built a protective wall 180 kilometres long to keep out refugees from the north. No finds relating to a period of three hundred years have been uncovered there; only afterwards did human settlement pick up again.[56] Core drilling in the Persian Gulf has shown that, at the time of the collapse, a terrible drought probably led to social and political problems.[57] A zone of drought, with significant water supply problems, must have reached from the Mediterranean as far as China. The 'Fertile Crescent' would have been ravaged by unbearably dry summers and short or nonexistent rainy seasons.[58]

In traditional societies, climatic turbulence and food emergencies call the legitimacy of rulers into question. The institutions holding responsibility – kings or priests – can react to a worsening of conditions only within the parameters of their culture. If their crisis-handling instruments are not sufficient, religious and political crises may appear alongside social and economic ones, resulting in the fall of a regime or the collapse of a civilization.[59] For agrarian civilizations a water supply failure is the worst case scenario, so it is scarcely surprising that not only the Akkadian and Egyptian kingdoms but the whole Mesopotamian civilization collapsed.[60] Against a background of alternating dry and rainy seasons, it seems logical that weather gods headed the pantheons throughout the region of Mesopotamia, Assyria, Babylonia, Mitanni, Hattusa, Ugarit, and so on. A quotation from the *Atrahasis* epic, in which the weather god Adad turns away from the people, illustrates the precision with which a lasting drought and associated salination of the soil could be observed. 'Above Adad made his rain scarce. Below the fountain of the deep was stopped, the flood rose not at the source. The field diminished its fertility; Nisaba [the grain] kept away, the dark fields became white, the wide unworked fields brought forth saltpetre.'[61]

Rise and fall of the river valley cultures

Around 2600 BC – almost simultaneously with Egypt – the civilizations of the Indus were in full bloom. Before then, around 3000 BC, the rainfall and vegetation had increased. Because of the high temperatures, the Indian river valley civilizations were heavily dependent on the quantity of rain; the reliable recurrence of the monsoon was favourable for the yield of early agriculture.[62] The Indus civilizations, nowadays grouped under the term 'Harappa', are characterized by regular, chessboard-like cities with a fortified acropolis and a fortified lower town consisting of brick houses and drainage systems.

According to Indologists, the end of the Indus civilization was due to an environmental catastrophe triggered by climate change. Archaeological finds point to a sudden drying up of the Ghaggar river around 1700 BC, which led to smaller harvests with devastating consequences for the cities. People disappeared along with their cities, and the Harappa civilization sank into oblivion.[63] Two centuries later, around 1500 BC, the area was resettled by nomadic herdsmen and horse-breeders, who came to southern Asia with a migratory wave of Indo-European peoples.[64]

Again there may be a counterpart to this decline in Egypt. For a similar catastrophe to that which overtook the Old Kingdom occurred there in the eighteenth century BC. Since the 11th Dynasty (c. 2134–1991 BC) Egypt had been passing through a new ascent; the 12th Dynasty counts as one of the high points of Egyptian civilization. During the Middle Kingdom (c. 2040–1650 BC) the annual Nile floods were less high than in the Old Kingdom, but they were constant and sufficed to produce high agricultural yields. This epoch reached its peak in the long reign of Amenemhet III (c. 1841–1797 BC). But then something comparable to the terminal chaos of the Old Kingdom set in. During the so-called 13th Dynasty, the kings came and went so often that it is still not possible to clarify their number and sequence, and in the subsequent 14th Dynasty the kingdom once again fell apart. On the basis of her excavations, Barbara Bell has pointed out that this second decline too was marked by Nile flood failure and resulting famine from 1768 BC on, and that the pharaohs suffered the same legitimation difficulties as in the earlier 'Intermediate Period'. She speaks of a 'Little Dark Age', which led to the breakdown of pharaonic rule.[65]

Europe's auspicious Bronze Age

The Bronze Age was a kind of golden age, in the course of which a series of major changes were brought into Europe's Neolithic world, into the world of the Etruscans, Thracians and other peoples of the third to the first millennium BC. Among the changes were the introduction of high-performance ploughshares, more intensive mining and long-distance commerce, newly specialized trades such as prospecting, mining, smelting, moulding, forging or bronze-trading, and the revolutionization of everyday life through more efficient tools.

54

Metal hammers, saws, blades, needles, and so on changed the production of countless objects such as leatherware and textiles. New industries arose to produce useable waggons and wheels, chariots and ships. A stronger vertical structuring of society corresponded to the differentiation of regional cultures, since bronze (an alloy of 90 per cent copper and 10 per cent tin) was expensive to produce.[66]

Conditions in the late third millennium BC were drier than at the postglacial climate optimum. The period of the *urnfield culture* was probably the driest since the end of the last great ice age, although the consequences north of the Alps were less serious than in the Mediterranean, North Africa or the Middle East. Irrigation difficulties must have reduced the quantity of arable land; high plateaux were abandoned; and people settled in river valleys or beside lakes. Moreover, forests were deliberately cleared, perhaps as a result of the drying up of fields used in agriculture. The water table lay considerably deeper than it is today. Hence the testimony to this Bronze Age culture has sunk in the literal sense of the word.[67]

The breakdown of civilization around 1200 BC and the beginning of the Iron Age

The first advanced civilization in the area of present-day Europe arose in Greece. Mycenae displays traces of settlement in the early Bronze Age (c. 2900–2500 BC). The flowering of this Bronze Age civilization began in the sixteenth century BC, more or less simultaneously with the New Kingdom in Egypt, and reached its peak in the fourteenth century BC. Mycenaean ceramics have been found, for example, in the palace of the Egyptian pharaoh Akhenaton (Amenhotep IV, r. 1364–1343 BC) in Amarna.

From the late thirteenth century BC Mycenaean civilization underwent a catastrophe of its own. The building of palaces ceased everywhere in Greece at that time. Around 1200 BC the castle buildings at Mycenae and most other noble residences in Greece were plundered and set on fire. Towns were abandoned, and in the following decades whole swathes of the interior remained deserted. The next few centuries are regarded as Greece's 'Dark Ages', because art, architecture and literature petered out and, until the beginning of the Homeric age four hundred years later, no written testimony lights up the darkness of history.[68]

The demise of Mycenae was traditionally linked to the Trojan war, which according to Homer's *Iliad* the Achaeans waged under Mycenaean leadership. This theory has always seemed implausible, however, since it was the Greeks who destroyed Troy, not vice versa. No evidence has been found in the Mycenaean castles for the alternative theory of a catastrophic earthquake. And, against the theory that a scarcity of bronze was responsible, there are the telling facts that the twelfth century BC was one of crisis everywhere in the Mediterranean and that there is no trace of any shortage of bronze.[69] Aristotle (384–322 BC) intimated that Mycenae must have dried up long before Homer's age, as Egypt did in his own time. Mycenae had been fertile at the time of the Trojan war but then turned into wasteland as a result of drought. 'In Argos', on the other hand, 'the . . . land that was formerly barren owing to the water has now become fruitful. Now the same process that has taken place in this small district must be supposed to be going on over whole countries and on a large scale.'[70] These statements sound clear enough. Yet only in the 1970s did anyone further develop this theory and argue that a persistent drought, in which Greece became dry and infertile, undermined the existential basis of Mycenaean civilization.[71]

Water shortage is also blamed for the fall of the Hittite empire, which occurred around 1200 BC after two hundred years of prosperity. Following terrible famines in Anatolia, the Hittites even asked Egypt for aid and shifted the centre of their empire from the highlands to the plains of Syria. But there they ran into another difficulty:[72] the famines in the Mediterranean unleashed the migration of 'sea peoples', the warlike raiders whose attacks led to the end of the Ugarit and probably also the Hittite empire.[73] It is interesting to note that for the Hittites the land belonged to the weather god, who merely entrusted it to the custody of the royal clan; the king's supreme religious duty was to enter into dialogue with the god.[74]

In the wake of the sea peoples' attacks, the nation of Israel took shape in Palestine after the decline of the old city-states.[75] Its god tolerated no other god beside him, and sections of the Jewish priesthood energetically combated the recurrent cult of Baal. Baal/Haddam was the traditional weather god in the region, whose status passed over into the Hebrew god: 'Then Moses stretched out his staff towards heaven, and the Lord sent thunder and hail, and fire came down on earth' (Exodus 9:23). Yahweh too, who first shows up in the list of place names compiled for the Egyptian pharaoh Amenhotep III (r. 1402–1364), was only another form of the Semitic weather god Haddam, though without the earlier attributes of bull, axe, thunderbolt or lightning flash.[76] The Hebrew deity is in the tradition of ancient Oriental weather gods, who punished humans with thunder and lightning, hail and tempest, flood and drought.[77] Christianity inherited the weather god along with Jewish monotheism: 'And the Lord will cause his majestic voice to be heard . . . with a cloudburst and tempest and hailstones' (Isaiah 30:30). The Old Testament is shot through with such images.[78]

The aridity of the Sub-Boreal affected Europe, North Africa and West Asia, but also other parts of the world. Thus, dendrochronological studies tell us that around 1200 BC the annual growth of Californian bristlecone pines began a sharp decline that lasted several centuries, which might point to a shift in heavy rainfall patterns. South Asia was also once more affected by drier conditions: monsoon crop yields must have fallen by 70 per cent in Rajasthan between 1300 and 900 BC, and pollen analysis allows us to detect the end of the ancient Indian civilization. It was during this period that the Great Indian Desert, the Thar, came into being.[79] In China the last decades of the Shang dynasty (c. 1766–1122 BC) witnessed climatic turbulence, a darkening of the sun by 'dry fog', the appearance of a triple sun, unnatural cold spells, frost in July, nighttime ice in the Yellow River valley (where it was normally much too hot), harvest failures and famines, heavy rainfall and flooding followed by a seven-year drought. These disturbances led to the fall of the dynasty and continued into the early years of the Chu dynasty (c. 1122–249 BC).[80]

The great upheavals around 1200 BC signified a far-reaching cultural shift. The rise of iron in the Middle East was perhaps due less to bronze shortages than to an increase in military conflicts. Unlike copper and tin, iron ore deposits were quite widely distributed. Anyone who had mastered the new technology could equip armies and win wars, and also produce cheap durable tools for craftsmen and farmers. The Iron Age ushered in the rise of new empires, which over the subsequent period absorbed many of the old commercial centres. Cities by no means lost their importance. However, the greater urbanization within the new empires was no longer geared to long-distance trade and local agriculture, but rather to extensive systems of tribute. The transition from Bronze Age to Iron Age is sometimes considered to be as significant as the Neolithic Revolution around 3000 BC.[81] And for some years it has been suggested, with good reason, that this cultural shift was associated with climatic change.[82]

The 'climate plunge' and political unrest around 800 BC

In rainy Europe the drier climate did not pose an insuperable problem; rather it was the temperature drop at the end of the Bronze

Age that caused major conflicts. Appropriately, Europe entered Iron Age culture at the same time. The long warm and dry period of the Bronze Age gave way around 800 BC – roughly 2,800 years ago – to the cooler climate of the ('post-warming') *Subatlantic Age*. The period is very roughly divided into two phases: Subatlantic I (the early 'post-warming' period, c. 800 BC to AD 1000), during which, apart from a mini-optimum phase in antiquity, it was rather cooler and wetter than today; and – after the medieval warm period – Subatlantic II ('later post-warming period', from c. AD 1300 to the twentieth century), in which we were living until recently. This phase of the Late Holocene coincided with the historical age of human civilization and will be considered more closely in the next chapters.

The 'climate plunge' around 800 BC was first discovered in archaeological excavations, then confirmed by palaeobiologists. Archaeologists found it especially interesting that different cultures prevailed in Central Europe before and after the climate plunge: the Bronze Age *urnfield culture* before (whose name comes from its burial rites, involving cremation and the keeping of ashes in urns); the Iron Age *Hallstatt culture* after. The link between climatic and cultural change is so striking that the question is often raised in the literature as to whether the transition to iron use was causally related to the less favourable climate. For the improved tilling of the soil with iron ploughs was able to offset declining agricultural yields, while iron weapons increased the chances of survival in an age when military conflicts were on the rise. This would seem to be an example of how technological and economic innovations ('progress') may be triggered by a worsening of the climate.[83]

The urnfield culture ended in a time of cooling and heavier rainfall. In many parts of Europe finds from the urnfield period lay beneath thick layers of mud; the Hallstatt finds are always higher, though only rarely in the same places. This means not only that the metal deposits and burial sites varied, but also that settlement patterns and ways of life changed. Palaeobiologists now ascribe different plant pollens to the archaeological strata and consider that nature too changed radically within a short space of time.[84] With the onset of the Subatlantic, average temperatures fell by 1 to 2 degrees Celsius, and rainfall increased quite appreciably.[85] Snow remained longer and over larger areas, glaciers grew, and tree lines moved lower (some 300 to 400 metres lower in the Alps, to the height at which they were at the beginning of the twentieth century). The high mountain pastures of the Bronze Age had to be abandoned, and Alpine settlements

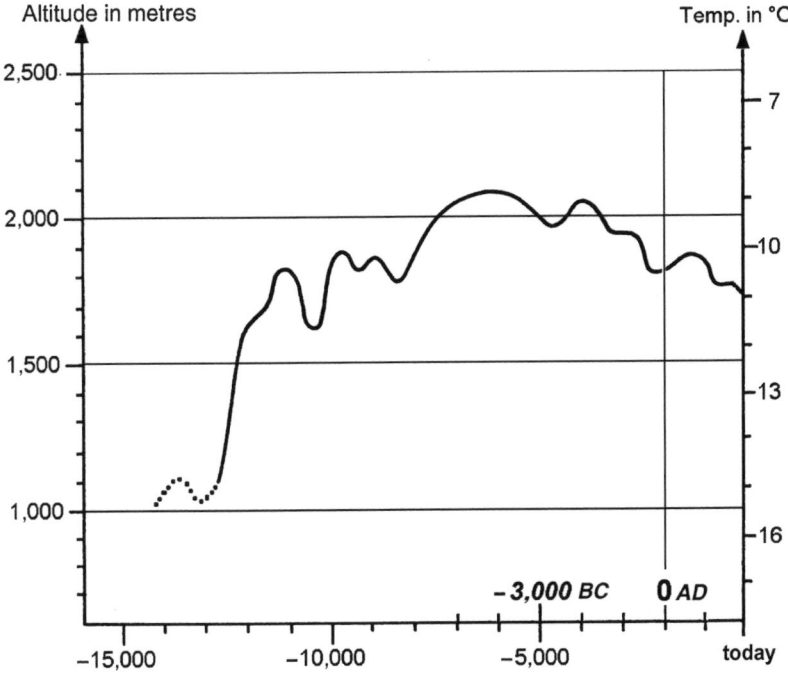

2.3 Changes in the tree line in Central Europe. The peaks are clearly visible in the Holocene maximum, as is the cooling around 800 BC and during the Little Ice Age.

thinned out. As water levels rose, lakeside locations became uninhabitable. Swollen rivers and newly inaccessible mountain passes made it necessary for people to change their travel habits.

Settlements were moved to higher, more secure altitudes; one example is the Magdalensberg near Hallstatt in Upper Austria. For the first time low mountain ranges, such as the previously water-poor Swabian Alb, were included in the areas of human habitation. The growing number of burial places in such altitudes is interpreted as a consequence of wetter conditions. At the same time, the food that people ate must have changed. The cereals grown in the Bronze Age were replaced with less demanding oats and rye; pollen analysis – for example, in the moorland around Lake Feder in Swabia – has shown a sharp discontinuity in this respect. We should probably assume that the conversion period witnessed undernourishment and a higher incidence of disease and mortality. It is also likely that the relative importance of livestock breeding vis-à-vis agriculture increased as a result of unreliable harvests.

The advance of salt-mining has also been associated with the climate change, since it was no longer possible to preserve food through wind-drying. If meat was to be kept, it had to be pickled in brine. This was all the more necessary because the Iron Age, with its farming improvements based on more durable ploughs, hoes and axes, had again brought about an increase in population. Salt mines such as the ones at Hallein-Dürrnberg and Hallstatt in the Salzkammergut experienced a boom that turned them into centres of culture. The earlier period of the Iron Age in Central Europe (up to the fifth century BC) is named, after them, the *Hallstatt period*.[86]

After what has been said so far, it is clear that the climate change around 800 BC – which in the literature has also been termed the '*Hallstatt disaster*'[87] – resulted in large-scale migratory movements and other adjustments to the new ecological conditions. Even in times of lower settlement density, the occupation of new land cannot have taken place without frictional losses. Furthermore, the growing importance of iron ore and salt mining, as well as the displacement of transport routes, led to the choice of different areas for settlement.

In other parts of the world, too, the 'climate plunge' around 800 BC resulted in migratory movements and warfare. In Egypt, after the reign of Pharaoh Takeloth II (860–835 BC), there was a political decline that led to collapse of the kingdom and persistent civil war. It was not previously thought that this anarchy was related to climate change. But, in the economic upturn that began in the seventh century BC, epigraphs plead for the beneficial effects of the Nile flood – which perhaps shows where the problems had lain until then.[88]

From Roman Optimum to Medieval Warm Period

The wet and cool climate of the Subatlantic period, with its cool summers and mild, rainy winters, lasted roughly until the time of Christ's birth – that is, through the whole age of the Roman kings and the Republic. The water table was probably higher than it is today, and in North Africa too the oases offered an ample basis for life. This explains why North Africa could become the 'granary' of the Roman empire.[89] The civilizations of the Greek and Etruscan city-states and of the Roman Republic developed in this favourable climate.[90]

To all appearances, the climate became warmer during the rule of the first Roman emperor, Augustus (63 BC to AD 14, r. 30 BC to AD 14). Temperatures were then probably similar to today's, or even higher north of the Alps. In the Near East and North Africa the greater aridity persisted. The Egyptian scholar Ptolemy (c. AD 100–160) compiled a weather diary around 120, whose regular reports on rainfall in each month of the year except August indicate significant differences from today's climate. Only in the fourth century did North Africa become drier.[91] Many settlements from this time were later swallowed up by the Syrian and Jordanian desert. For the period around the birth of Christ, we should probably operate with a figure of 300 million for the population of the world; already roughly a half lived in the two ancient civilizations of Asia, China (80 million) and India (75 million), with 35 million each in West Asia and Europe and another 15 million in North Africa. So, during the Roman optimum, more people lived on our planet than at any time before. This level would be reached again only a thousand years later, during the warm period of the High Middle Ages.[92]

The heyday of the Roman empire

Of course, a world-historical phenomenon such as the rise of Rome from Italian city-state to world power cannot be seen as the result of climate change. Not only did the process unfold over too long a period, but many other factors played a role. Rome's ascent took place at the cost of the Etruscans, Greeks and Phoenicians, who lived under the same climatic conditions. At most it might be said that, whereas Carthage, the great power of the southern Mediterranean, reached its high point during the colder phase, Rome had its golden age after the warming began, when the political centre of gravity had shifted to the northern side of the Mediterranean. But, even after the destruction of Carthage in 146 BC, the Roman province of *Africa* remained for centuries one of the economically most important provinces of the *Imperium Romanum*. Perhaps it is significant that Rome initially expanded southward and struck out north only after the warming had begun.

With Octavian's elevation to Augustus, the growing empire acquired a monarchical head who took the structural changes into account through the unification of law, the systematic exploitation of subject areas, and the development of a coherent foreign policy. Under Trajan (c. 53–117, r. 98–117) the *Imperium Romanum* attained its maximum extent: it stretched from the borders of

Scotland to the Caspian Sea and the Persian Gulf. This great expansion coincided with a quite warm but not excessively dry period, which is known in climate history as the *Roman climatic optimum*.[93] The warming, which lasted from the first century until approximately AD 400 and – because of glacier melting – corresponded to rising sea levels, probably helped to consolidate this first northern Mediterranean empire or, in any event, its northward expansion. The fact that passes over the Alps were clear all year round made it easier to conquer and control the transalpine provinces of Gallia, Belgica, Germania, Raetia and Noricum.

In the high Alps, mining could be conducted in regions where permanent frost was still the rule at the end of the twentieth century.[94] In the writings on horticulture by Pliny the Younger, it is stated that wine and olives were cultivated in more northerly parts of Italy than in previous centuries. An edict of Emperor Domitian (51–96, r. 81–96) prohibited the extension of viticulture to provinces north of the Alps – which can only mean that efforts were being made in that direction. An edict of Emperor Probus (232–82, r. 276–82) rescinded this order in AD 280, and viticulture was introduced into Germany and Britain with such success that there are scarcely any more reports of wine imports from the south after the year 300.[95]

Formation of the Eurasian empires

The ancient climate optimum favoured the creation of empires from Europe through the Middle East to East Asia. Political stability helped to promote long-distance trade. The Chinese civilization built the first major empire, under the despotic ruler Ch'in Shih Huang-ti (r. 246–210 BC), who passed into the afterlife with his famous army of terracotta warriors. His empire was shattered by a popular uprising, but a new dynasty operated pragmatic policies that succeeded in restoring stability. The observation that Han dynasty China (202 BC to AD 220) flourished at almost exactly the same time as the Roman empire is not new. China, like the West, had its formative period in classical antiquity. The Han Chinese are still today the main constituent of the People's Republic of China, accounting for 90 per cent of the population (98 per cent in Taiwan). In the Han dynasty, despite its financial crises due to military expenditure, China experienced a unique economic boom. Census figures put its population in AD 2 at approximately 60 million.[96]

Of course, the climate not only helped the empire to prosper but also stimulated peoples in the north to greater activity. As new areas of settlement developed, the population expanded in both size and energy in the northern regions. The second and third centuries saw the beginning of the great migrations of Goths, Gepidae and Vandals, who moved into southern Russia and the Carpathians. The north of Europe pressed downward. The Roman empire tried to stem this advance from the north by building fortification systems such as the Limes in Germany and Hadrian's Wall in Britain.

In the area of today's Mongolia, the 'Xiongnu' tribes forced their way into the Chinese empire. The Great Wall was designed to keep them at bay, and in the second century the Xiongnu were steered towards the West. They advanced towards India and eventually to the Black Sea. In Europe, the geographer Ptolemy first described them and gave them the name '*Chunni*' or '*Hunni*': that is, 'Huns'. In 376 they crushed the East Gothic empire of King Ermanarich in southern Russia, then routed the West Gothic army of Athanarich and drove the Burgundians and Vandals out of eastern Europe. The victory of the Huns triggered the so-called migration period, the influx of Germanic peoples into the Roman empire.[97]

Decline of the great empires

It would be tempting to relate the crises of the Roman empire and Han China to climatic developments, if there were not so many other factors in play such as their susceptibility to external attack, the pressure to militarize society and the mounting fiscal burden. Initially these were factors which, though arising from a structural crisis of the empire, could be mastered by a capable ruler such as Marcus Aurelius (121–80, r. 161–80). But under Emperor Commodus (161–92, r. 180–92) they were compounded by famine and epidemics, gangsterism and conspiracies. In 189 his deputy was murdered in protests due to food shortages.[98] The next hundred years saw no fewer than forty emperors. Under the Barracks Emperors (235–85) the empire came to the brink of the precipice, and under Decius and Valerian the first systematic persecutions of Christians began. After the thousand-year celebrations of the founding of Rome, Decius (c. 190–251, r. 249–51) fell in combat against the Goths, and Valerian (c. 200–62, r. 253–60) ended his days in Persian captivity. Under his successor, Gallienus, the empire broke up into several parts, as famine and outbreaks of plague led to population decline and a partial return to natural economy. Emperor Aurelian (c. 214–75, r. 270–5) elevated the 'invincible sun god' (*Sol invictus*) to the official imperial deity. It needs to be investigated to what extent the crisis of the third century correlates with climatic influences.[99]

After the *Imperium Romanum* had recovered politically and economically, Theodosius (347–95, r. 379–95) achieved a new unification of the empire. It should be noted that climate historians consider conditions to have been favourable at the time. The Roman optimum made another appearance, with its warm but not excessively dry climate. In 395 Theodosius thought he could stabilize his legacy by dividing the empire into a Western and Eastern (Byzantium) half. But the fifth century was much less auspicious: the climate grew cooler, and Rome's traditional granary in North Africa dried up. In 410 the western Goths sacked Rome, then the Vandals roamed right across the western provinces before coming to rest in Roman North Africa in 429. The Burgundians settled in Savoy in 443, the Franks in the Lower Rhine area, and the Alemanni in the Upper Rhine. The fifth-century crisis of the empire took on such dimensions that people were scarcely surprised when the last West Roman emperor Romulus Augustus (r. 475–6) was deposed by a Germanic army commander.

Eugippius (c. 465–533) describes the collapse of the Roman empire in his account of the life of St Severinus of Noricum (d. 482), an ascetic who interpreted it as God's punishment for the sins of humanity.[100] Although bad weather seemed one of the lesser evils in a world of wars, expulsion and violence, Eugippius constantly refers to cold, famine and disease. Severinus organized dispatches of aid, but the goods-laden ships were iced up on the River Inn. His own prayers and the acts of atonement performed by believers led God to order a thaw, and the starving people received food galore. Throughout the ordeal the holy man not only lived sinless but chastised himself with fasts and flagellation. The severest trial was in matters of clothing: 'So at midwinter, which in those regions is a time of cruel, numbing cold, he gave a remarkable proof of endurance by being always willing to walk barefoot. A well-known proof of the terrible cold is afforded by the Danube, which is often so solidly frozen by the fierce frost that it affords a secure crossing even for carts.'[101]

In northern Europe cold was the main problem; in the Near East, North Africa and parts of Asia it was aridity. During periods of drought the level of the Caspian Sea fell to a minimum. In southern Italy, Greece, Anatolia and Palestine, people moved to the coasts and left the hinterland largely uninhabited. This was the period when great cities went into decline: Ephesus, Antioch and Palmyra in Asia Minor. Some six hundred settlements were abandoned in Arabia, where elaborate irrigation systems had previously kept agriculture going. The expansion of the Arabs, and the associated spread of Islam, ensued at a time of adverse climatic conditions in their traditional homelands.[102]

The Chinese Han empire declined at the same time as the Roman empire. Strife in the imperial household and disputed legacies certainly played a role in this, as did religiously inspired popular insurrections. As in Rome at the time of the Barracks Emperors, the military took over political responsibility in China. In AD 220 warlords divided up the empire (the 'three empires period'). The decline was aggravated by significant cooling, aridity, harvest failure and famine. The Yangzi froze over more than once, and in years of drought the great rivers almost dried up several times. In 309 it was possible to cross the bed of the Yellow River or the Yangzi without getting your feet wet. Nature's hostility to life, combined with government incompetence, provoked disturbances and uprisings.[103] The pressure from peoples to the north increased with China's internal weakness. The Chinese period of mass migrations began, and decline set in during the period when the empire was divided (220–589). Again as in the West, salvationist religions grew amid the turmoil and loss of power, pushing into the background both Taoism and the philosophy of Confucianism (which lost its foundations with the decomposition of the empire). For half a millennium Buddhism became the leading spiritual force: it promised consolation in harsh times and encouraged passivity by orienting people to the afterlife. At the same time, a new written culture blossomed in the monasteries. Because of these parallels, the fall of the Han empire is sometimes spoken of as the beginning of the 'Chinese Middle Ages'.[104]

The catastrophes of the Early Middle Ages

Although there are again disagreements over dates, most authors concur that the climate worsened in late antiquity. According to Helmut Jäger, the colder winters and wetter climate had already set in by AD 250 and lasted in the north until 750 and in the main part of Europe until the ninth century. The cooling of this *Early Medieval Pessimum* averaged 1–1.5 degrees Celsius throughout the year. The glaciers grew, and tree lines fell by some two hundred metres in Central Europe. In higher locations, or in more northerly regions, conditions deteriorated for the growing of wine and cereals. There was a rise in harvest failures and susceptibility to disease. Mortality increased among new-born babies, infants and old people.[105]

Schönwiese situates the Early Medieval Pessimum from AD 450 to 750, and points out that in central England a lasting rise in temperatures, by 1 to 2 degrees Celsius, only set in around the year 1000.[106] Hubert Horace Lamb (1913–97) thought that the summers tended to be warmer and drier until around AD 400, that periods with a colder and changeable climate become discernible only in the fifth century.

In the northern Mediterranean and in northern, western and central Europe, the cold periods were associated with wetter conditions. We have reports of stormy weather and floods, which led to coastline changes on the North Sea and in southern England. In Italy the incidence of flooding suddenly shot up in the second half of the sixth century.[107] In Switzerland the glaciers – for example, the Lower Grindelwald glacier – assumed the dimensions they would have later towards the end of the Little Ice Age, making impassable an old Roman road through the Val de Bagnes. Lamb concluded that it may not have been only political threats that harried the Roman administration during the period of its breakdown in the West.[108]

One of the classical stereotypes is the idea that the 'migration period' which led to the end of the empire in the West had climatic causes; the worsening climate of late antiquity or the Early Middle Ages is accordingly termed the *Migration Period Pessimum*.[109] A simple way of refuting this causal link is to point out that the migration period stretched over centuries. In fact, there were different reasons during different phases. The real migration began under the conditions of the climatic optimum, possibly as a result of major demographic growth in the North. The great migration period was triggered in AD 375 when the Huns broke out of the Asian steppes. They precipitated the westward migration of the Germanic peoples, who advanced into the territory of the Roman empire. Their occupation and settlement of land led to the destruction of the West Roman empire. Byzantium was able to reconquer Italy from the Ostrogoths and North Africa from the Vandals. But Spain remained Visigoth ('Catalonia') or Vandal ('Andalusia'), Gaul Frankish ('France'), Britain Anglian ('England') and Raetia Allemanic ('Allemagne'). The mass migration ended in 568 with the conquest of Italy by the Longobards, who were able to settle there in large numbers ('Lombardy').[110]

The breakdown of the empire was accompanied with a demographic implosion: by the sixth century only half of its 15 million inhabitants were left. Especially in the former Latin parts of the empire (Pannonia, Germania, Gallia, Hispania, Britannia), but also in North Africa, the population figures fell sharply as a result of external attack, wars, harvest failures and epidemics, as well as outward migration. This decline caused settlements, roads and farmland to fall into disrepair. Cultivated areas were reconquered by nature.

The complete abandonment of settlements, for which there is considerable archaeological and palaeobotanical evidence, cannot be explained only by wars. Good places to live are normally preserved even when there is a complete change of inhabitants. In late antiquity, however, most settlements north of the Alps were given up. Only a few centres exhibit even a weak continuity of habitation into the Middle Ages. Pollen analysis testifies to a general decline of agriculture. The forests marched forward and engulfed human settlements in the space of a few decades. The land would look like virgin nature to future generations who came to occupy it again. The location of new villages and the development of new settlement patterns in the sixth and seventh centuries point to changed climatic conditions that made the cultural hiatus necessary.[111]

The volcanic eruption at Rabaul in March 536 and the Plague of Justinian

The Byzantine historiographer Procopius of Caesarea (c. 500–562), who lived in Rome, wrote that throughout the tenth year of Emperor Justinian's rule (482–565, r. 527–65) the sun lacked brightness and looked more like the moon. Lydus of Constantinople also tells us that the sun was dull during this same year, when army commander Belisarius was at the height of his fame, and that the fruits of the field withered at a quite unusual time. Zacharias of Mytilene confirms that the sun was obscured by day and the moon by night, from the 24th of March in year VIII (of the fifteen-year *indictio* cycle) until the 24th of June in the following year. And John of Ephesus reports from Asia Minor that the darkening lasted eighteen months, and that the sun was visible for a maximum of four hours a day; the crops therefore failed to ripen and the wine was sour. In his ecclesiastical history the same author adds that the winter of 536–7 was exceptionally severe, with heavy snowfall in Mesopotamia. Apparently these extreme climatic conditions affected both the Middle East and Europe.

The historical account of atmospheric phenomena induced vulcanologists to search through the climate archives. Ice core drilled in Greenland does indeed show a strong acidic signal, reckoned in independent samples for c. AD 540 (plus or minus ten years) and c. AD

535. As no comparable evidence of volcanic fallout can be detected in the broad surrounding area, these signals can refer only to the volcano-related disturbances recorded by Procopius. Richard B. Stothers has identified a volcano in the southern hemisphere as the cause of this mysterious cloud – most likely Rabaul in Papua-New Guinea. Radio carbon measurements on the spot have dated its eruption to the time around AD 540 (±90 years), and there was no other such contemporaneous event anywhere in the world. If we assume that the dispersal of its ash and aerosol was similar to that of Tambora in Indonesia, then the volcano must have erupted two to three weeks before, in early March 536. Rabaul's acidic signal in the Greenland ice is twice as high as Tambora's in 1815, and so we may reckon that the worldwide effects were also twice as great.[112] Perhaps a connection should be made between this darkening of the skies and the famines of the late 530s and the Plague of Justinian.

An age of insecurity in Europe

The early Middle Ages were a time of extreme insecurity in Europe: the population level fell to a low that was never again reached at any later period; harvest yields and storage capacities were greatly reduced, and crop failures and famines were a frequent occurrence. The annals scrupulously record the cold spells and floods, harvest failures and famines, human and livestock epidemics. If we take the accounts left by Bishop Gregory of Tours (c. 538–94), the picture in Frankish Gaul and Visigothic Aquitaine for the 580s – a decade when he was writing from his own experience – is one of persistently heavy rainfall and stormy weather, thick snows and late frosts from which even the birds died, serious floods, mountain avalanches, livestock epidemics, harvest failures, famine and epidemics both familiar and unfamiliar.[113] Whole areas were depopulated, and the survivors had to endure the breakdown of all infrastructure. With reference to the early Middle Ages, the French medievalist Georges Duby speaks of a 'hostile environment' and 'a long period of cold and wetness'.[114]

The threat of harvest failure, famine and epidemic was more serious than that of wars. In the Alps the glaciers advanced from the early fifth to the mid-eighth century – a sure sign of general cooling.[115] Nature inspired fear in men and women, its wildness was compared to chaos. Wolves fell upon herds and travellers. In the lean year 843, a starving wolf burst into a church in Sénonais while the service was taking place. No holds were barred in the fight against ravenous beasts: traps, poison baits and organized hunts each played a part,

and Charlemagne (AD 747–814, r. 768–814) ordered the employment of special wolf hunters in every county of his kingdom.[116] Frosty
winters, spring floods and summer droughts caused harvest failure
and famine. In another hungry year, 784, a third of the population
was said to have died. People baked bread from every unfit substance
imaginable; in Saxony they ate horse meat, and there were even isolated cases of cannibalism. Chronic undernourishment was one of the
factors behind the high mortality. For the period between 793 and
880, despite the paucity of sources, Pierre Riché established that there
were thirteen years of famine, thirteen with floods, nine with epidemics and another nine with extreme cold in winter.[117]

Wet summers and hailstorms were especially damaging to crops.
In a society with no concept of the accidental, there was a tendency
to personalize misfortune. Archbishop Agobard of Lyons (769–840),
a man of Visigoth origin, wrote in his sermon *De grandine et tonitruis*
(On Hail and Thunder): 'In these parts nearly everyone – nobles and
common folk, town and country, young and old – believe that human
beings can bring about hail and thunder. . . . We have seen and heard
how most people are gripped by such nonsense, indeed possessed by
such stupidity, that they think and say there is a land called Magonia.
Ships are supposed to have come from there in the clouds, and to
have taken back the crops destroyed by hail or ruined in stormy
weather; the sailors in the sky gave the weathermakers a payment
and received cereals and other crops in return.'[118]

As the climate worsened, it affected not only the size of harvests
but also the quantity of livestock. The carcass weight of pigs or cattle
was lower than that of today or of the Roman period. Floods led to
livestock disease, which in turn fuelled extreme xenophobia among
farmers. Agobard of Lyons wrote of the epidemics of 810 in France:
'Just a few years ago a wave of stupidity spread because of the death
of a livestock animal. It was said that, because Duke Grimald of
Benevent had fallen out with the Christian emperor Charles, he sent
people here with a powder and told them to sprinkle it over fields,
hills, pastures and springs; and that it was from this sprinkled powder
that the animal died. We have seen and heard how many were impris-
oned and some put to death for this reason; most of the latter were
tied to boards, thrown in the river and killed.'[119]

If it was a long winter, farm animals could no longer be fed. Crops
withered in a dry summer and rotted in a wet one. Each harvest
failure resulted in famine for the people. In the imperial annals we
read the following for the reign of Louis the Pious (778–840, r.
814–40): 'This year, persistent rainfall and an inordinate amount of

moisture in the air led to great hardships. People and animals alike have been ravaged everywhere by such epidemics that there is hardly a district in all of France which appears to have been spared. Even grain and vegetables were ruined by constant rains and either could not be harvested or rotted in the storehouses. The grapes did not fare much better; the harvest was meagre, and the wine tasted harsh and sour because of a lack of heat. In some areas it was impossible to take care of fall planting because of flooding of the plains by rivers, so that not a single kernel of grain got into the soil prior to spring.'[120] This period also saw the triumphant progress of leprosy, the characteristic deficiency disease of medieval Europe.[121]

The rise and collapse of Maya civilization

With regard to the European Middle Ages, we must consider to what extent its climate system also appeared in other parts of the world. After all, the period from 650 to 850 is presented in accounts of China and Japan as a warm phase.[122] The Indian Maya civilization of Central America, which had developed since AD 300 and flourished during the climate pessimum in Europe, may serve as a perfect example. Classical Maya civilization, with its city-states featuring tall pyramids and great architectural complexes, had a dense population, a highly developed priestly caste, a distinctive script and astonishing knowledge of astronomy and mathematics. It rested upon a productive agriculture, which today's world has to thank for some of its most important crops: potato, maize, tomato, avocado and tobacco. Yet this civilization suddenly collapsed. No new buildings or inscribed monuments were put in place after the year 900. The population declined dramatically as cities and other settlements were abandoned and cultural knowledge was lost.[123]

Those who have investigated the possible causes, apart from war, point to chronic overpopulation, environmental destruction through clearances and overuse, and resulting harvest failures, famines, epidemics and rebellions directed against the upper layer of Maya society. The Maya did not die out as an ethnicity, but the nobility and priesthood disappeared from the scene.[124] After the demographic collapse, patterns of settlement changed. The post-classical Maya no longer settled in the highlands, but lived instead by the shores of seas, rivers and lakes. This has fuelled the idea that water supply problems caused the fall of classical city-states such as Chichén Itza, Palenque or Tikal. The archaeologist Richardson Gill has shown that the period with least water of the past seven thousand years was between the years AD 800 and 1000. This aridity robbed civilization of its existential basis, and the population collapsed amid crises due to famine. Wars and rebellions were the final straw.[125] A group of geologists, who worked on an ocean drilling programme off the coast of Venezuela, have been able to show from varve sedimentation that in the centuries of the Maya collapse the summer monsoon was permanently drifting south and the necessary rains failed to appear in Mexico.[126]

The threat of drought explains the great importance that Maya culture attached to worship of the rain god, Chac. For the period of collapse, four extreme droughts come into consideration as crisis triggers: one of several years around AD 760, one of nine years around 810, one of three years around 860 and one of six years around 910. These dates fit what we know from the Maya glyphs: namely, that the first group of cities in the western lowlands around Palenque collapsed about the year 810, when there was scarcely any access to the groundwater reservoir; that fifty years later the urban cultures in the southeastern lowlands around Copán followed suit; and that about the year 910 the same fate befell the cities of the central and northern lowlands around Tikal, Uxmal and Chichén Itza.[127] The lowland karst holes, the *cenotes*, kept water longer. The topography of the peninsula, together with climate change, would thus explain the temporal gradation of the collapse. Meanwhile the fall of Maya civilization has been re-enacted in climate models.[128]

The succession of civilizations in the Peruvian lowlands and highlands

The South American counterpart is the Moche civilization in Peru, which began its rise around 100 BC and reached a climax in the sixth century. Early in the seventh century, buildings were destroyed and acts of war were committed in the heart of the empire. A little later, the Moche civilization collapsed. There is much to suggest that, in the coastal strips that are today so arid, it was increased rainfall that led then to a cultural blossoming. Water reservoirs, canals and aqueducts encouraged more intensive agriculture and served to minimize risks to the environment. Great pyramids, built with adobe bricks at the mouth of the River Moche (Huaca del Sol, Huaca de la Luna), were the main holy sites of the empire.

Discussion of the reasons for the collapse has focused on the usual suspects: external invasion, civil war, overpopulation, harvest failure, famine and climatic change. Excavations covering the period around AD 600 have revealed traces of flood-like precipitation, which would have led to the destruction of settlements, major damage to the central holy sites and probably also to areas of arable land. The disastrous floods were followed by decades of greatly reduced rainfall. Persistent drought led to harvest failures, famines and violent conflicts, to a decline in land areas under cultivation, and to rapid changes in the flora and fauna. Excavations at the holy sites tell us that, at the time of the disasters and cultural decline, hundreds of people must have been ritually sacrificed to calm the wrath of the gods. These facts, combined with ice core analysis from the Quelccaya glacier in the Andes, show that great aridity prevailed for centuries on the Peruvian coast. More recently, major El Niño events have been held responsible for heavy rainfall.[129]

Already in the 1970s, the archaeologist Allison C. Paulsen established that prosperity alternated between coastal and highland cultures in what are now Peru and Ecuador. After the decline of the Moche civilization, the coastal areas of present-day Ecuador were deserted because of extremely arid conditions, whereas in the southern Andean highlands a civilization known as the Huari Tiahuanaco empire flourished. Then, after its decline in the ninth century, new civilizations arose on the southern and northern coasts of Peru, which reached their apogee during the European high Middle Ages. After lack of rainfall in the fourteenth century dissolved these in turn, the best-known Andean civilization, the Inca empire, began its ascent.[130] While its artefacts have been dated with radiocarbon methods, annual ring counting of ice core in the Quelccaya glacier has yielded evidence of similarly dramatic climate change. It has become clear, for example, that a long drought hit the southern highlands between 563 and 594, and this must have affected the living conditions of people in the area. The fact that there was no civilization in the highlands at that time is scarcely surprising. A group of climatologists has suggested synchronizing the emergence of civilizations with the precipitation record in ice core samples.[131]

Remote effects of the 'Christchild'

The alternation of drought in the southern highlands and heavy rainfall in the northern highlands put climatologists on the scent. Precisely this alternation, with short-term variations, marks a regional climatic phenomenon that is known as El Niño ('the Christchild') because it appears in the Christmas period. Warm water covers the cold deep waters of the Humboldt Current, reducing the size of fish catches. This leads to heavy rainfall in the normally dry Peruvian coastlands, while rain fails to materialize in the southern highlands (as in the big 'El Niños' of 1975–6 and 1982–3). A stronger El Niño appears every three to seven years, accompanied with a general change in ocean currents and in atmospheric conditions in the South Pacific. This effect on currents is known to oceanographers as the '*El Niño Southern Oscillation*', or ENSO for short. Occasionally it begins not in December but only in summer, when it is longer-lasting. In these 'super El Niños' there is a stronger interaction between atmospheric pressure and the winds. Then changes in precipitation levels appear throughout the tropics: heavy rainfall in otherwise dry areas, periods of drought and forest fires in otherwise wet areas. ENSO events are the most important natural climatic signal on a world scale. In the view of some climatologists, however, there is also a kind of 'mega El Niño', in which the climatic anomalies stretch over decades. Then precipitation levels can be above average on the northern coasts of Peru and Ecuador, while arid conditions prevail in the southern highlands.

It is interesting that 'super El Niños' affect not only the east coast of South America but also the whole southern part of the northern hemisphere. In the wake of ENSO events, the usual climate turns everywhere into its opposite. When warm water flows on the east coast of South America, large parts of the continent become warmer and, in many cases, wetter. On the other hand, conditions become drier in the tropical north of Australia, in Oceania, Indonesia and the Philippines, and in the Indian subcontinent. The monsoon winds necessary for agriculture are weaker or non-existent. In Madagascar and East Africa it is drier and hotter than usual, but in equatorial Africa and the south of North America it is wetter and the winter is colder. In the north of North America and in Japan the winters become warmer. Climatologists think that even 'mega El Niños' can be detected in the Peruvian ice core. And, because of long-distance effects, the South American data can be correlated with findings from ice core in Asia. This has been studied with the help of drilling in the Dunde glacier of Tibet. The synchronicity of anomalies may mean that we are dealing here with common causes.

Comparative analysis of ice core from Peru and China has shown that their glacier growth-rates had similar tendencies for nearly four hundred years, between 1610 and 1980, even though the Dunde glacier lies north and the Quelccaya glacier south of the Equator.[132]

As long ago as the seventh century, ENSO events and the level of the Nile flood were linked to each other through rainfall in East Africa. By the early twentieth century it was understood that drought years in India and Australia were also years of low rainfall in Egypt, whereas high Nile floods were observable in years with a strong monsoon. Records of the 'Nilometer' near Cairo may therefore be significant for all regions affected by El Niño – and these records go back five thousand years, having been kept with great accuracy since late antiquity. What they show is that ENSO events were most frequent in the early Middle Ages, with a high point around the year 800. They were also common and strong during the so-called Little Ice Age (c. 1300–1900). But they did not make themselves felt much during the medieval warm period. The significance of this discovery, and what it means for the global environment, have not yet been fully discussed.[133]

The high medieval warm period (c. 1000–1300)

The idea of a *medieval warm period* was formulated in 1965 by Hubert Lamb, who based his conclusions on historical texts and physical data regarding the climate. He located its peak between AD 1000 and 1300, in the high Middle Ages.[134] During this time there was a whole series of warm dry summers and mild winters. Lamb estimated the warming at one to two degrees Celsius above the average in the 'normal period' 1931–60. In the far north it was as much as four degrees warmer. There are scarcely any reports of drift-ice between Iceland and Greenland. Excavations in Greenland were made in places where permafrost still prevailed in the late twentieth century.[135]

Advocates of the 'hockey stick' (see above, figure 0.2) look askance on the theory of a medieval warm period, because it allegedly serves to downplay the anthropogenic warming of the late twentieth century. If, without any human influence, the climate in the twelfth century was even warmer than at the height of industrialization, why should not today's warming also have 'natural' grounds? Lamb's originally rather jejune estimate of a two degree warming (based on proxy data) thus became a matter for disapproval, because it far exceeded the 0.6 degree Celsius measurement for warming in the twentieth century. For this reason attempts have been made to shatter the credibility of the medieval warm period as such.[136] Raymond Bradley and his fellow campaigners – Michael Mann, for example, inventor of the hockey stick theory – would like to discuss the medieval warm period

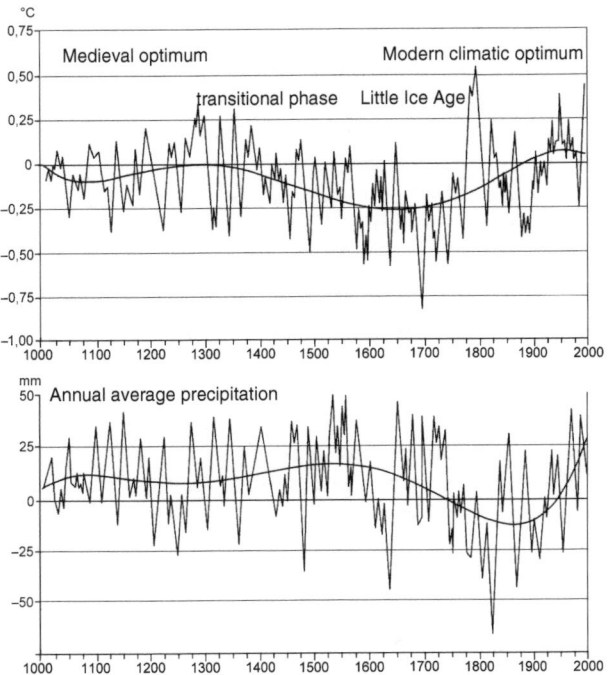

2.4 Short-term and long-term temperature variations and precipitation levels over the last 1000 years in Central Europe, according to detailed research by the geographer Rüdiger Glaser.

out of existence and, basing themselves on hydrological anomalies in North America, use instead the concept of a *medieval climatic anomaly*.[137]

A warm period in the high Middle Ages can scarcely be disputed, however, if we stick to the climate data and compare them with the turbulence of the early Middle Ages or the subsequent Little Ice Age. First of all, it has been shown that the large glaciers retreated in the period between 900 and 1250/1300, not only in Europe and North America but all around the world.[138] Pierre Alexandre, a French climate historian who has not been drawn into Lamb's terminology, has concluded from an extensive evaluation of European sources that no reliable records exist before about 1170. In general he notes wide regional differences – for instance, between countries north of the Alps and the Mediterranean area. In his view, the high Middle

Ages – as is usually the case in warm periods – were marked by heavier rainfall and a seasonally differentiated evolution of temperatures. The winters were actually rather colder, whereas the spring temperatures important for vegetation were three degrees warmer between 1170 and 1310 than in the climatic period from 1891–1960, which also saw the glaciers retreat. After the end of the warm period there was a cooling that reached a peak in the 1340s. In the Mediterranean a period of great aridity can be detected between 1200 and 1310.[139]

Two American isotope specialists who have studied the intensity of solar radiation take a new position in the dispute surrounding the medieval warm period. They have demonstrated that such a period existed in the high Middle Ages, without undermining the theory that our contemporary global warming is anthropogenic. Indeed, in their view the medieval warm period is proof of the anthropogenic nature of today's phenomenon. In arguing this, they assume that increased solar activity in the twentieth century can account for no more than a 0.2 to 0.4 degree warming; the remaining 0.2 to 0.4 degrees Celsius *must* therefore be caused by mankind.[140]

Heat years and capricious weather in Europe

The high medieval warm period also saw harsh climatic extremes, such as the winter of 1010–11, when the Bosphorus froze over and ice floated on the Nile.[141] In the winter of 1118 drift-ice was spotted off Iceland, frosts continued into June in Saxony, and freezing conditions gripped the lagoon in Venice: it was possible to ride across ice into the centre of the Republic of St Mark. The next summer was marked by hunger and widespread deaths. It took more than a hundred years before the lagoons and the Po froze again, as late as 1234.[142] The 1180s, on the other hand, were the decade with the warmest known winters; in January 1187 the trees blossomed in Strassburg. There had been warm or hot periods before then, such as the one between 1021 and 1040. Nuremberg sources complain that in 1022 people 'languished and stifled on the streets because of the great heat'; streams and rivers, lakes and springs dried up; water was in short supply. The summer of 1130 was so dry that it was possible to wade across the Rhine, and in 1135 the Danube had so little water that people could cross it on foot. Some authorities were shrewd enough to exploit the low water levels: for example, the foundations were laid in the same year for the famous stone bridge in Regensburg.

Hot dry summers were the rule from the 1180s in a remarkably long warm phase, until a cold stormy season in 1251 put an end to them and led to harvest failure, price rises, famine and major outbreaks of disease. Nor were warm summers auspicious, as Central Europe experienced drought and forest fires in years of great dryness. The longest phases of persistent summer heat occurred in Central Europe between 1261 and 1310, and again between 1321 and 1400. Biblical plagues, in the form of locust swarms, spread northward – in August 1338, for example, through Austria, Bohemia, Bavaria and Swabia to Thuringia and Hesse. In the case of many plants, it was not only the summer weather but even more the conditions in spring and at harvest-time that were decisive. Surprisingly, the springs were quite varied during the warm period of the high Middle Ages. Cool or cold temperatures alternated with moderate, warm or hot, although not as commonly as in the Little Ice Age. It must be asked to what extent the chronicler's viewpoint distorted things. Negative events were usually recorded. The warm, often dry autumnal weather in the second half of the thirteenth century features in relatively few reports, whereas adverse conditions at harvest-time (for example, the freezing of vines as early as 9 September in Alsace) receive a lot of attention. Temperatures seem on average to have been below those of the previous century. One has the impression that unfavourable harvest periods became more frequent in the early fourteenth century, at the onset of the Little Ice Age.[143]

Southern plants and insects in northern latitudes

Findings for the high Middle Ages are unambiguous about cultivation limits. The tree line in the Alps rose to over two thousand metres, a level not attained even in the Bronze Age optimum and far higher than in the twentieth century.[144] The tree line is also a pointer to shifts in the whole ecosystem: the trees were preceded in their upward movement by lichens and moss, grasses and flowers, together with their corresponding insects, birds and little mammals. Field names show that in the German high Middle Ages vines were cultivated not only in the old Roman areas on the Main, Rhine and Mosel, at locations two hundred metres above present-day levels, but much wider afield in Pomerania and East Prussia, as well as in England, southern Scotland and southern Norway.[145] Viticulture indicates that night frosts were rare and that there was sufficient sunshine in summer.[146] In the twentieth century, the Saale-Unstrut region marked the northern limit in Germany – and its wines were until recently more for people with offbeat tastes. But, in the new century, quality vine-growing areas have begun to develop in Mecklenburg and Belgium – and still the northern medieval limits have not yet been reached.

Pollen analysis has shown that Norway had crops in the high Middle Ages that would disappear with the onset of cooling. Wheat was grown as far north as Trondheim, and strains of barley up to a latitude of nearly 70 degrees. The agricultural area with a settled population of farmers began to expand in the ninth and tenth centuries, and by the height of the warm period it had advanced on average 100 to 200 metres further up the hillsides. The greater part of this cultivated land was lost after 1300.[147] In many areas of Britain, agriculture extended into uplands that had not been cultivated for a long time before, nor would be subsequently. In the desolate hills of Northumberland, for example, the arable land stretched as high as 320 metres above sea level.

In Asia, too, plants migrated northward. We know from old Chinese records that the cultivation limits of citrus fruits and Chinese grass (*Boehmeria nivea*) have never been as northerly as they were in the thirteenth century. Both are subtropical plants, whose yield closely depends on adequate heat. In 1264 their cultivation limits were several hundred kilometres further north than in the twentieth century.[148]

The study of fauna – for example, excavations in York, the Roman Eboracum – also reveals more favourable temperatures at that time. York lies slightly further north than Moscow, but thanks to the Gulf Stream it enjoys a relatively mild climate. The nettle groundbug (*Heterogaster urticae*) here serves as a climate pointer. There is archaeological evidence that it was present in York during the Roman optimum, and again during the Norman high Middle Ages. In the Anglo-Saxon and Viking periods, however, and again in the Little Ice Age, it was completely absent. And, despite the moderate warming, the nettle groundbug can still be found only in sunny areas of southern England. In the built-up medieval heart of York, traces of the *Aglenus brunneus* beetle also point to high temperatures during the medieval warm period.[149] The larger habitat for heat-seeking insects also had implications for the spread of diseases. The *Anopheles* mosquito was present in many parts of Europe, and malaria was accordingly endemic as far north as England during the high Middle Ages. Swarms of locusts, which today we associate with Africa, repeatedly caused crop failures in Central Europe as late as the fourteenth century.[150]

The end of famine and the blossoming of European civilization

The high Middle Ages led to the ebbing of famines and to long-term social progress. The population already increased in the ninth century, even doubled by some accounts.[151] And harvest failures did not reverse the trend. There were key improvements in agriculture, not only in comparison with the early Middle Ages but even vis-à-vis Roman antiquity. Thus, the horse collar and the head yoke for ox ploughs were first introduced in the eleventh century, in both cases representing an advance in animal traction. The heavy wheel plough spread quickly, and the introduction of the harrow promoted more rational cultivation of the soil. Shoeing made animals less susceptible to accidents. Crop diversification reduced the danger of harvest failure, and the introduction of pod vegetables (peas, beans, lentils) kept the ordinary population considerably better supplied with proteins and carbohydrates. The alternation of crops also helped to prevent soil exhaustion.[152]

If we were to put it in modern terms, we might speak of a booming economy. This was expressed in the use of new technologies in industry and agriculture, the two often being closely interlinked; one thinks, for example, of the cultivation of flax and dye plants, which supplied raw materials for the new urban weaving industry. The textile crafts that now began to take shape in the cities introduced more rational techniques such as the spinning wheel or the horizontal loom. Whole new trades appeared, such as paper production, for the rationalization of which paper mills were soon being built. In the twelfth century windmills came to join the existing watermills. The church and the nobility profited from the growing number of bondsmen and taxpayers to invest in the construction of fortresses and castles, churches and monasteries. This was necessary, since the small Romanesque churches could no longer hold the growing population.

The dawning of the new age led to a new architectural style. Early medieval churches typically had massive walls and tiny windows; they were dark and musty inside. The new Gothic style was both brighter and lighter. Huge windows allowed the sunlight of the medieval warm period to stream into the monumental cathedrals that were being built around this time. New inventions such as the wheelbarrow, the screw jack and the hydraulic saw were used to execute these gigantic projects. Forest clearance had made wood available in abundance. Huge numbers of trunks were used for scaffolding as well as for timbering in the great cathedral naves. The building work required new quarries to be opened and transport capacities to be created, and iron was needed for tools and equipment. Together with the demand for armour, weapons and military equipment, this stimulated iron ore mining, the iron industry and inter-regional trade in iron.[153] For the first time Europe attained a rank comparable to that of other advanced civilizations.

Population growth, forest clearance and settlement density

Europe's population grew by leaps and bounds, reaching dimensions it had never seen before. Settlement activity did not stop at political or cultural boundaries: the high Middle Ages was a time of new farmland and village creation even in remote areas and borderlands, even in low mountainous regions, high Alpine valleys and fjords. The Cistercian order, founded in the twelfth century, actually specialized in land clearances, farming and rural crafts. The medieval clearances took forms that made deep inroads into nature. Primeval forests that had arisen in the early Middle Ages were largely removed. It is estimated that the area covered by forest in Central Europe shrank from 90 per cent to 20 per cent (compared with 30 per cent today).[154] High mountain pastures were opened up as the warming made it possible to drive livestock up and to keep them there for longer periods. Moors began to be cultivated, and dykes were built by the North Sea, not only to reclaim land but also to prevent land loss. For there can be no doubt that the melting of glaciers in the higher temperatures raised sea levels. Parts of the Halligen islands were swept away by storm floods. Jadebusen bay on Germany's northwestern coast was formed in 1219 and Dollart bay in 1277–87. But these land losses were marginal in comparison with the arable land reclaimed from the sea.[155]

At this time Europe developed its characteristic landscape with high settlement density, in which remnants of forest and islands lay amid huge areas of farmland divided up into regular, relatively small parcels. Medieval German place names often allude to the forest (*-grün, -wald, -hain, -schwanden or -schwendi*) or to clearances (*-scheid, -schlag, -au, -stock, -reut, -roth, -rode, -rade*) and burnings (*-loh, -brand* and *–bronn*). Similar etymologies exist in other European countries: typical suffixes for cleared land are -thwaite and -toft in England or -tuit and -tot in France. What nearly all the new places had in common was a less favourable location than that of earlier settlements. Charts of the Black Forest make it clear that settlements in the ninth to twelfth centuries were often on marginal lands, either too deep in side valleys or too high above sea level.[156] At the end of the high medieval warm period that would prove to be a disadvantage.

At first, however, population figures rapidly increased. Of course there are no reliable statistics for this period, but demographic historians basing themselves on numerous pointers are agreed about the population growth in Europe. Some 46 million people lived there in 1050, 50 million a century later, 61 million in 1200, and no fewer than 73 million at the end of expansion around 1300. These people needed food, clothing, housing and spiritual care, which could no longer be supplied in sufficient quantity in the villages.[157] The high Middle Ages was therefore the great age of town formation. In German-speaking areas, the number of towns and cities grew from the few hundred that had survived the cultural decline of the early Middle Ages to more than three thousand; the figures look much the same in other European countries. In this urbanization drive, Europe reached today's number of major centres; towns that would later become successful big cities were founded at this time. New towns added to the total in later centuries remained marginal. Although the roots of European urban life lie in antiquity, the present density of distribution goes back to the age of the high medieval climate optimum.[158]

The beginning of European expansion

Only in the high Middle Ages did Europeans become sufficiently confident to embark on expansion abroad. This was, of course, not simply a response to climate change but also required a cultural motivation. As on the Islamic side, they were fighting for their God.

After the onset of the warm period, reconquest of the Iberian peninsula from Arab invaders was the first attempt at expansion. Spurred on by successes in this campaign, Pope Urban II (c. 1035–99, r. 1088–99) called in Clermont in 1095 for a great crusade, a 'holy war' against Islam. Following years of battles in Asia Minor, the French and Normans who formed the First Crusade succeeded in capturing Jerusalem in the year 1099.

Viking expansion and the formation of states in northern Europe

The high Middle Ages was not least the age of the Vikings. In the mid-ninth century they conquered areas in England; Norwegian

Vikings settled in the Shetlands, Orkneys and Hebrides and, year after year, fell upon the Celtic kingdoms of Scotland, Ireland and Wales. The 'Norsemen' founded states in Ireland, conquered 'Normandy' in western France in 911, and built a kingdom in Sicily. Vikings set up kings over the Slavs, beginning with Rurik (r. 862–79), who founded the dynasty of Kievan Rus that would rule the Russian empire down to the time of Ivan the Terrible (1530–84).

Scandinavia itself saw the founding of powerful kingdoms. Norway was unified under King Harald Fairhair (c. 850–c. 933, r. 860/70–c. 933) and, as in a chain reaction, Sweden followed under Bjorn the Elder (r. c. 900–50) and Denmark under Harald Bluetooth (c. 910–86, r. 950–86). The favourable climate of the high Middle Ages enabled the formation of Nordic nations that are still in existence today. Around the year 865 the Norwegian farmer Floke Vilgardson risked settling on a large island in the North Sea. When all his livestock perished in an extremely cold winter, he bid farewell to the island and gave it the terrible name: Iceland. But only nine years later Ingolf Arnarson landed there and enjoyed greater success. The medieval warm period began earlier in northern than in central Europe. The 'iceland' now proved an attractive country in which to settle, as even there agriculture and livestock breeding were possible in the tenth century. Within the space of just two generations, from 870 to 930, the Icelandic 'land grab' spawned a population of sixty thousand. Whereas the peasantry elsewhere in Europe was subject to feudal lords, the Icelanders had their freedom. Even before their conversion to Christianity – agreed at the Allting, the national assembly, in the year 1000 – they settled the icy north, as we know from the *Land-namabok* (or 'Book of Settlement') and the large number of pagan graves.[159]

The medieval warm period was Iceland's best age, but even then life was not easy. Great swathes of the country, and everywhere above 500 metres, were covered with large glaciers, which cut off the inhabited coasts and valleys from one another. Nor did volcanic activity bring with it only the agreeable experience of hot springs: one of the major disasters was the eruption of Hekla in 1104, which laid waste the densely populated Thjorsa valley in the south. Subsequently Iceland went through 'a millennium of misery', marked by harvest failures, famines, epidemics and a dramatic fall in population and livestock numbers. The eleventh-century total of 80,000 inhabitants was not achieved again until the twentieth century. In the late eighteenth century, after the eruption of Lakikagar, the Danish authorities of the day discussed a complete evacuation of the island.[160]

The settlement of Greenland

Vikings led by Erik the Red (c. 950–1005) sailed west from Iceland to a much larger, deserted country. When Erik returned three years later with a view to tempting settlers there, he called it 'the green land'. In 985 he set off with a fleet of twenty-five ships carrying new settlers, seed corn and livestock. The term 'Greenland' struck later chroniclers as rather suspect, and Erik was reproached with self-serving optimism or deception. For, already in the twelfth century, the far north of the island cooled again. But in Erik's lifetime the appellation might have been more applicable. When the warm period was at its height in the far north, it was even possible to grow cereals in Greenland, and the ship route from Norway to Iceland and Greenland was ice-free all the year round. Climatologists are of the view that storms were uncommon during this period, so that the fabled seaworthiness of the Vikings – before the invention of the compass and with no comforts on board – was due not least to the favours of the climate.[161]

In Greenland, Erik founded two areas of lasting settlement. The one near the southern tip (Österbygd, or 'eastern settlement'), where the leader himself built his farm, became the island's political centre; recent excavations have uncovered about 450 farms, far more than are mentioned in the written sources. The other one (Vesterbygd, or 'western settlement') actually lay far to the north, but on the western side facing America. Throughout the high Middle Ages there was regular sea traffic between Norway and both these areas of settlement. After it was Christianized, Greenland was even included in the Roman episcopal organization in the early twelfth century; it was allocated to a bishop whose see was at Gardar, on Eiriksfjord. The Viking graves on the island are in an area where permafrost prevailed in the twentieth century, though evidently not at the time of the burials. Should the soil thaw out again in the twenty-first century, it will thus return to its state during the high medieval warm period.

As the Icelandic sagas record, new lands to the west were discovered from Greenland when ships were driven off course by a storm. This led to subsequent journeys of exploration. The land in the west, today's Labrador in Canada, which was then given the name *Markland* ('land of forest'), filled the explorers with enthusiasm. For trees did not grow in the harsh climates of Iceland and Greenland, where wood had to be imported with difficulty from Norway. Leif the Fortunate (c. 975–1020), the son of Erik the Red, discovered

Helluland, today's Baffin Island, and sailed from there southward to another island that he called *Vinland* ('vineland'). About 1005, Vikings under Thorfinn Karlsefni even began to settle in North America. The sagas record this, and we also have the evidence of the *Anse aux Meadows* site in Newfoundland excavated in the 1960s. From this settlement numbering more than one hundred inhabitants, further voyages were made to the south. But the hostility of the Skrælings, as the Vikings called native Americans, led to the collapse of the first European colony in the Americas. The sea routes from Greenland, not to speak of Iceland or Norway, were too distant to lend it support.[162]

— 3 —

GLOBAL COOLING: THE LITTLE ICE AGE

The Concept of the Little Ice Age

The term 'Little Ice Age' was coined in the late 1930s by the US glaciologist François Matthes (1875–1949). It first appeared in a report on recent glacier advances in North America,[1] then in the title of an essay on the geological interpretation of glacier moraines in the Yosemite valley. Matthes was interested in the coolings since the postglacial climatic optimum, that is, over the last three thousand years, and especially in that which followed the medieval warm period. In his view, most glaciers still existing in North America do not go back to the last great ice age but have arisen in this relatively short space of time. The period from the thirteenth to the nineteenth century, in which glaciers advanced in the Alps, Scandinavia and North America, he called 'the Little Ice Age' (to distinguish it from the great ice ages).[2]

The term was taken up in 1955 by the Swedish economic historian Gustaf Utterström (1911–85), who suggested a worse climate as the explanation for economic and demographic problems that Scandinavia faced in the sixteenth and seventeenth centuries.[3] His essay pointed to inadequacies in the argument of Fernand Braudel, the leading social historian of the time, who focused only on social factors. Eric Hobsbawm similarly concentrated on social-economic factors, interpreting the political crises up to the English Revolution as a history of class struggles.[4] Nevertheless, this interest in 'the crisis of the seventeenth century' drew fresh attention to the early modern period, whose centre was no longer sought in the Reformation or the French Revolution but rather in the crisis-ridden middle of the period.[5]

Rejecting the Durkheimian principle observed by social historians, whereby the social is explained by the social,[6] Utterström emphasized instead factors from outside the system of society: that is, the climate. The French historian Emmanuel Le Roy Ladurie branded this essay an extreme example of the 'traditional method' (which he did not define more precisely) in the historical sciences.[7] Yet, when he set about refuting Utterström, he noted that the climate theory fitted quite well the structural-historical interests of the Annales School,[8] and that he himself had been infected with it. Ladurie took up serial analysis of wine harvest data and, a few years later, published a standard work on climate history that crucially rested upon Utterström's proposals, drew heavily on scientific literature and made the Little Ice Age more widely known.[9]

Since then the concept has aroused growing interest. In a number of major research projects, Hubert Horace Lamb (1913–97) in England,[10] Christian Pfister in Switzerland,[11] Rudolf Bradzil in the Czech Republic[12] and Rüdiger Glaser in Germany[13] have so clearly demonstrated climatic fluctuations in European history that their existence is now beyond all doubt.[14] As in the case of the high medieval warm period, however, it is necessary to begin by mentioning that we have to do here not with constant cooling but with a dominant tendency. Along with a large number of cold and wet years there were also periods of 'normal' weather and even some years of extreme heat. Many climate historians today cautiously define the period of the Little Ice Age by climatic variability and frequency of extreme events.[15]

Analysis of two weather diaries from Bavaria (Ingolstadt and Eichstätt) for 1508 to 1531 has shown that winter temperatures during the Reformation period were similar to those in the 1970s. On the other hand, for the decades after 1563, a clear fall of 2 degrees Celsius in average temperatures is observable in the Zurich area – a cooling event typical of the Little Ice Age.[16] In a medieval perspective, a distinct climatic period may be said to have begun around the year 1300 that had little in common with the warm high Middle Ages; and in the second decade of the fourteenth century began the Little Ice Age.[17]

Possible causes of the global cooling

The causes of the Little Ice Age are uncertain, first of all because of the paucity of sources.[18] A *decline in solar activity* is usually regarded as the main explanation for the global cooling. Observers in China,

Japan and Korea, working within a tradition that went back to late antiquity, recorded the absence of sunspots in the late seventeenth century. In Europe, global cooling theories appeared more or less simultaneously with the invention of the telescope, the building of observatories and the systematic study of nature. The lesser number of sunspots was interpreted as a sign of reduced solar activity and linked to the cold spell that began in 1675. The cold phase between 1675 and 1715 is called the *Maunder Minimum*, after the astronomer Edward Walter Maunder (1851–1928), who collected and summarized observations made at the time.[19] The solar activity hypothesis is one of the most elegant explanations, both for the Little Ice Age as a whole and for its extreme years calculated by George C. Reid with the help of a climate model.[20]

In the 1970s a group of Danish geophysicists around Claus U. Hammer pointed out that the ice record indicates two clusters of intensified *volcanism* that coincide with the high point of the Little Ice Age. In contrast to the sparse sulfate signal for the high Middle Ages, they found traces of the strongest volcanic activity since antiquity for the period between 1250 and 1500 and again between 1550 and 1700.[21] Our understanding of particular eruptions is only gradually increasing. Thus, analysis of 33 ice cores in Greenland and the Antarctic has dated to the year 1452–3 the explosion of the volcano Kuwae on Vanuatu, which preceded the 'global chill' of the 1450s.[22] The eruption must have been more powerful than that of Tambora, but it caused major cooling only in the southern hemisphere.[23]

Five volcanic eruptions have been identified for the precarious decades between 1580 and 1600; they too had an influence on European history, even if most historians have never heard of them. These were the eruptions of the Billy Mitchell volcano at Bougainville (Melanesia) in 1580, of Kelut on the island of Java in 1586, of Raung on Java in 1593, of Ruiz in Colombia in 1593, and of Huaynaputina in southern Peru in 1600.[24] Only the last of these is also known to us from extant sources, kept by the Spanish colonial authorities, and we can date it precisely to 19 February 1600. It devastated large areas within a radius of twenty kilometres, sent ash raining down on Peru, Bolivia and Chile,[25] and entered the stratosphere with such effect that solar radiation was reduced worldwide over the following months.[26] Harvest failures and famines occurred in the global wake of this event. More generally, climatologists have worked out that eight periods with especially cool summers are linked to eight major volcanic explosions.[27]

The Changing Environment

Worldwide glacier growth and increased aridity

The grand old lady of glacier research, Jean Grove (1927–2001), to whom we owe the first book specifically on the Little Ice Age,[28] emphasized in her review of ice core records from around the world that a climatic optimum in the high Middle Ages was followed by several centuries in which waves of cooling led to new glacier advances. This was a worldwide phenomenon, even if it was not always completely synchronous. Grove placed the onset of global cooling in the early fourteenth century, and even a few decades before that in the far north.[29] The high glacier level lasted, with ups and downs, until the end of the nineteenth century.[30]

It is true that the global cooling did not everywhere lead to glacier growth, and so the term 'Little Ice Age', developed in glacier research, sounds somewhat inappropriate for arid and tropical regions. In West Africa and similar regions near the Equator, it was not so much the cooling as the irregularity of rainfall that posed a threat. As in parts of the Mediterranean, drought was the main problem. The climate zone in which agriculture could be pursued became dramatically smaller during the early modern period, because the Sahara desert and the Sahel zone moved several hundred kilometres south. The old centre of Timbuctou on the upper reaches of the Niger, which in 1600 had still been an agriculturally productive region on the northern fringes of the savannah, lay two centuries later on the edge of the Sahel.[31]

In India and tropical Central America, too, the shift in seasonal rainfall represented the greatest problem. Increased *aridity* may be regarded as the typical feature of the global cooling. Spain dried up.[32] Venetian officers reported long periods of drought between 1548 and 1648 on the island of Crete. In 25 per cent of the years, not a drop of rain fell all winter or in spring; it was devastating for the fields and the grape or olive harvests. (In the twentieth century there is no such example of a complete absence of rain.) A fifth of all winters, on the other hand, were marked by 'exceptional falls of snow, protracted periods of abnormal cold, or rain so excessive as to prevent sowing of crops until late spring'.[33]

Admittedly, what may be a real find for geologists and glaciologists throws up methodological problems for historians; whereas the former range over relatively long climatic periods, the latter operate with smaller chunks of time. At the level of years or months it is much more difficult to prove a worsening of the climate. Something which may appear in geological terms as a single process of cooling, because glacier formation and sedimentation take place over decades or centuries, is more difficult to demonstrate at the level of daily life. For people living at the time, short-term changes had greater importance than medium to long-term ones. So we find in their records more observations about the weather than about the climate; they were attentive to very cold spells, long winters with much snow or persistently icy conditions.[34] Local chroniclers did make comparisons over longer periods, and priests referred to the monstrous masses of snow for their spiritual exhortations.[35] In 1624, after one extremely long and hard winter, the Thuringian clergyman Martin Pezold (d. 1633) even put together a chronicle of European extreme winters in his book of *Schneegedancken* or 'snow thoughts'.[36]

The growth of glaciers was carefully recorded. In 1601 the peasants of Chamonix turned in panic to the government of Savoy, because

3.1 The Little Ice Age in Matthäus Merian's copper engraving. The growth of the lower Grindelwald glacier threatens traditional settlements and becomes a tourist sight.

the glacier known today as the *Mer de glace* was growing larger and larger, had already engulfed two villages and was about to destroy a third.[37] Martin Zeiller (1589–1661) wrote in Matthäus Merian's *Topografia Helvetiae* of the Grindelwald glacier near Interlaken in the Bernese Oberland: 'Not far from town there used to be a chapel to St Petronel, to which people made pilgrimage in times of old. Since then the mountain's tendency to grow has covered the place. So the local people watch and notice that the mountain is growing hugely and driving the ground or earth before it, so that where there used to be a fine meadow or pasture it is disappearing and turning into raw, desolate mountainside. Indeed, in several places houses and huts along with the peasants living in them have had to move away because of its growth. Also growing out of it are big rough ice floes, as well as rocks and whole pieces of cliff, which thrust aside and upward the houses, trees and other things present there.' After an account of the glacier's movement and the thundering crash of thawing ice floes, the author ends by noting that the mountain's growth is conjuring away 'the peasant's pasture, commons and houses. It is therefore a truly miraculous mountain.'[38]

The freezing of lakes, rivers and seas

It has been concluded from the total freezing of major lakes in China that the average temperature there between 1470 and 1850 must have been one degree colder than in the late twentieth century.[39] The number of *Seegfrörne* – the Swiss German word for the freezing of large Alpine lakes has become known beyond the country's border – was significantly higher in fifteenth and sixteenth-century Europe than ever before or since. Lake Constance develops a full ice cover only when the temperature is −20 degrees Celsius or lower for a protracted period of time. We have chronicle reports of its freezing over as early as the ninth century, in the winters of 875 and 895. Then the lake was free of ice for nearly two hundred years. In the eleventh century there were two *Seegfrörnen*, in the twelfth Lake Constance froze only in 1108, and in the thirteenth total icing over occurred three times.

The frequency increased, however, with the first cold waves of the Little Ice Age. In the fourteenth century *Seegfrörne* were reported in 1323, 1325, 1378, 1379 and 1383. A peak was then reached in the fifteenth and sixteenth centuries, with seven instances in each. Between 1409 and 1573 the lake was on average completely frozen over every twelfth year, and in the very heart of the Little Ice Age, between 1560 and 1575, every fifth year. The longest incidence was probably in the winter of 1572–3, when the lake froze in December, briefly thawed at Epiphany (costing a few people their lives), then froze even more solidly than before. Only on Easter Monday, 24 March, did the ice melt again. During the time of complete ice cover, the nearby communities took possession of the lake: they used it for excursions, distance measurement, smuggling, skating trips and carnival festivities, as well as for regular trade in goods. Carts with teams of six horses even dared to cross the lake.

On 17 February 1573, the '*ice procession*' that has become a tradition down to the present day first took place. On that occasion, people from Münsterlingen in Switzerland carried a bust of St John to Hagnau in Swabia (where it would remain until the next *Seegfrörne*), and then the procession made its way back. The lake froze over twice in the seventeenth century: in 1684 and 1695, two of the coldest years of the millennium. In the century of the Enlightenment there was only one *Seegfrörne*, in 1788, and in the nineteenth century a total of two, in 1830 and 1880; we have numerous lithographs and photographs of these events. In the twentieth century the only instance of complete ice cover was in 1963. Even ice-breakers were unable to clear a way for ships, and as in the worst years of the Little Ice Age frozen birds had to be collected in droves.[40] Climatologists of the time thought this was the beginning of a new ice age. But the bust of St John has been waiting ever since in Münsterlingen for the lake to freeze again.

Other major Alpine lakes, such as Lake Geneva, Lake Lucerne and Lake Zurich, have not always frozen in the same year – which points to the existence of microclimates. But the dates we have for them confirm the trends just described.

The situation was similar for great European rivers such as the Rhine or the Thames. In the 1560s it was more than once reported from Cologne that the Rhine had not only frozen but iced up right down to the river bed. The frost *fairs on the Thames* became famous in the sixteenth to eighteenth centuries: as soon as the ice could bear the weight, life in the capital city of London moved onto the river, complete with trading booths and winter sports. Even eating houses were set up there with open fires. The scene was so spectacular that it has been handed down to us in numerous woodcuts, copperplate engravings and oil paintings. From the Netherlands too – as famous as England for its normally mild winters – we have countless pictures from the Little Ice Age that show people skating or playing hockey on frozen canals and rivers. Rivers even froze in the Mediterranean: the Po near Venice, the Arno near Florence, the Rhone in southern France or the Guadalquivir in southern Spain. In January 1709 the Saone succumbed near Lyons: it froze all the way down to its bed, and heavy rainfall in the previous period meant that the ground also froze to a depth of three feet (approximately one metre). That same winter it was not only the olives, vines and sweet chestnut trees that froze: wine was not safe in cellars, nor the ink in inkwells. Livestock froze in their stalls, as did wild animals in the forest. Birds fell dead to the ground. The Mediterranean froze over in the Bay of Marseilles.[41] Such cold winters were characteristic of the Little Ice Age.[42]

More lasting, however, were the changes in the northern seas. Here the drift ice and pack ice reached much further south. The north of Iceland – ice-free all year round both today and in the high Middle Ages – became cut off by ice during the long half-year of winter. Harbours in the south, such as Reykjavik, remained free only for a few months. The harbour of Spitzbergen, which today, as in the high Middle Ages, can be reached by ship nine months of the year, was open only during the three summer months in the Little Ice Age. The winter of 1315–16 was so cold that the Baltic froze over – a spectacle often repeated in succeeding centuries. The drift ice limit extended so far south in the fifteenth century that shipping to Greenland, and at times even to Iceland, was interrupted. Icebergs hindered the passage of ships to Norway, and even to Denmark or the British Isles.[43]

At Venice the lagoon froze twice in the twentieth century: once for four days (10 to 13 February 1929) because of persistently icy winds (the Bora) with a temperature of −10 degrees Celsius; and once for eleven days (10 to 21 February 1956), when the Bora winds measured −8 degrees. This was colder than during the four centuries of the high medieval warm period, when good sources inform us that the lagoon froze over only twice. On the other hand, we know that during the Little Ice Age, from 1300 to 1800, this happened no fewer than thirty times: six per century. The first occurrences came with the onset of the Little Ice Age, in 1311 and 1323. Sometimes the results were quite spectacular: for nearly seven weeks in 1432, from 6 January to 22 February, it was possible to travel by carriage from Venice to Mestre on the mainland. In the winter of 1491 a knightly tournament was held on the ice of the Grand Canal. In 1569 the lagoon froze again in March, later than ever before. In the extreme winters of 1684 and 1709 the ice supported the transport of heavy goods. Venetian diaries report solid ice cover in 1716, 1740, 1747 and twice in 1755. A number of paintings exist of the freeze in 1789.[44]

3.2 The Thames often froze during the Little Ice Age, and markets and sporting events took place on the London ice. The last ice cover, in 1895, consisted only of floes pressed together.

Changing flora and fauna

The Protestant pastor Daniel Schaller wrote in the late sixteenth century: 'In towns and villages, much weeping and moaning is heard among farming folk that the land has grown weary and even exhausted of bearing fruit, causing great price rises and famine.'[45] Such voices should be taken seriously as indications of how people at the time perceived the effects of climate change on flora and fauna. For, as at the beginning of the Iron Age, quality cereals proved susceptible to wet conditions and winter cold. In some regions of northern Europe – Iceland, for example – the cultivation of cereals had to be abandoned altogether; in others wheat was given up and people made do with oats and rye. From harvest observations we know that fruit blossom time, the haymaking season or the grape ripening period was delayed as a result of bad weather conditions. In some years, summer

93

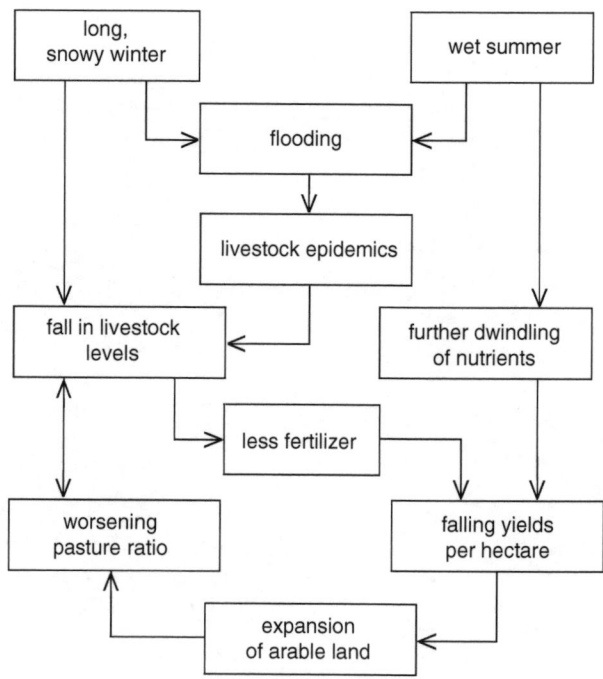

3.3 Effects of the Little Ice Age on agriculture. Hypothetical model by Christian Pfister (Berne).

north of the Alps was too short for the grapes to ripen, or only sour wine was produced.[46]

The shifting cultivation limits affected two core sectors of European agriculture: wheat and vines. Whereas in the high Middle Ages viticulture had even existed in southern Norway and England, the limits moved quite a long way south in the fourteenth century and again in the sixteenth. By Pastor Schaller's time, viticulture had completely dropped out of the picture in the Baltic. And still today, after the warming of the past century, its limit is some five hundred kilometres further south than at the end of the high Middle Ages.[47] Even in the most favourable locations on the Rhine and Mosel, where wine-growing has held up in Germany, the wine itself was in some years undrinkable.[48] In October 1588 the Cologne patrician Hermann Weinsberg (1518–97) was astonished to find himself running out of wine, because it had been of such poor quality in the previous thirteen years that he had been unable to lay in fresh supplies. In a jocular allusion to his own name, he sighed: 'So now the Weinsberg [*Weinberg* = vineyard] has to celebrate without wine.'[49]

We do not yet know for sure what happened to the whole range of economically useful plants, but Fernand Braudel already established that the cultivation limit for olives shifted south.[50] Our knowledge about changes in the ecosystem is defective mainly because we lack systematic observations about wild plants. It is clear that the tree line fell in the high Alps and that high mountain pastures had to be abandoned. There has also been some speculation that the range of herbs contracted in Alpine meadows, with consequences for cattle health, milk production and the quality of dairy products.[51] In extreme years and in unfavourable places – high-lying valleys and low-mountain regions, for example – plant diversity probably declined. Further investigation would be required to determine the extent to which the tree stock also changed.[52]

People living at the time noted some of the effects of the climate change on animal life: 'The waters are no longer as rich in fish as they once were, the forests and fields no longer as abounding in game and animals, the air not as full of birds.'[53] The reference to fish here probably reflected the real situation. We actually know more about saltwater fish, because records of catches in the North Atlantic tell us that cod stocks off Iceland and Norway have sharply declined since the late Middle Ages. This is explicable in physiological terms: cod livers stop functioning below 2 degrees Celsius and may thus serve as a gauge of temperature. In extremely cold years in the seventeenth century – 1625 and 1629, for example – or during the Maunder Minimum, the fishing limit shifted further south and no cod could be caught even off the Faroe Islands.[54]

On land the evidence is more impressionistic. For example, the extinction of the bearded vulture (*Gypaetus barbatus*) in the late sixteenth century has been linked to an absence of thermal activity in the Alps.[55] In connection with the 'indescribable cold' and the freezing of Alpine lakes in the winter of 1570–1, Johann Jacob Wick (1522–88) reported 'very deep snow, and how many people froze and suffocated to death in the snow'. As in the early Middle Ages, ravenous wolves came out of the forest and attacked human beings: Pastor Tobias Egli (1534–74) from Chur reported to the Swiss reformer Heinrich Bullinger (1504–75) one such case involving three seamstresses near Zizers in the Rhine valley.[56] As to game animals, harvest failure and famine brought a twofold danger: they suffered from the shorter periods of vegetation, and they were exposed to

3.4 In the extreme winter of 1570–1 wolves came out of the forest. Illustration of the snow disaster in the collection of reports by Johann Jakob Wick (Zurich).

more intensive hunting. In times of great privation, the criminal sanctions against poaching became quite drastic. Around the year 1600 hunting bans applied to many species of birds in the Bernese Oberland, because their numbers had fallen to dangerously low levels.[57] We can see from account books that, in some years of extreme cold, drought or flooding, the mole population must have fallen so dramatically that bounty payments no longer had to be paid.[58]

With regard to fauna too, what we know about farms is again greater than about the wild. In some countries of northern Europe – Greenland and Iceland, parts of Scandinavia and Britain – cattle-raising became impossible and farmers switched to sheep. In low-mountain regions and in the Alps, loss of pasture reduced the scope for livestock-breeding. The greatest danger came from snowmelt flooding, since it poisoned pastureland and led to animal epidemics.

Newe Zeyttungen.
Oder/Warhaffte anzeigung/wie schröcklich die grosse Erd
bibem/Sturmwind/vnd grosses Regen wetter/so sich zum theil zu end desi abgelauffnen
1612. zum theil aber auch in dem Monat Januarii/ dises sent louffenden 1613. Jars/ an vilerley Orten
erzeigt/ vnd was merckliche schaden dardurch verursacht worden.

3.5 Insurance company statistics can provide no information about past natural disasters. Woodcut depicting the great 'Thuringia Flood' of 1612–13.

The latter were a major theme in the agrarian societies of the early modern age, as we can see from the surviving sources.

The climate change also affected insects and micro-organisms.[59] Cooling north of the Alps was bad news for the *Anopheles mosquito*, and it put paid to the malaria problem. Whereas in the high Middle Ages malaria had been present even in England, it retreated to North Africa during the Little Ice Age (today it is becoming endemic in Iraq and south of the Sahara). On the other hand, conditions in the North improved for many other parasites. *Fleas* and *lice* felt especially good in thicker clothing where personal hygiene was deficient – as the term *'clothes lice'* suggests. Lice transmit the *Rickettsia prowazekli* bacterium, the cause of an especially dangerous form of typhus (*Typhus exanthemicus*), which even today is fatal in 10–20 per cent of untreated cases.[60] The *Yersinia Pestis* flea was the carrier of plague,

97

which had been endemic in Europe since the mid-fourteenth century, and the new cooling helped to spread it more widely The decline of bathing culture in the sixteenth century – as nudity and sexuality became taboo – also improved living conditions for the flea. This explains the literary interest in fleas, examples of which are the satirical play *Flöh Haz, Weiber Traz*,[61] by Johann Fischart (1546–90) and the macaronic verse of one Witzbold, who called himself *Knickknackius* and described his native North Germany as 'flealand'.[62]

The end of the Vikings in Greenland

The Nordic countries were hardest hit by the cooling. In Greenland the increasingly adverse conditions put an end to European settlement. The vegetation period grew dramatically shorter, cereal-growing became impossible, grazing land shrank, and wood imports from North America or Norway became more and more difficult until they finally ceased altogether. Trade with the mother country broke off. We can therefore only conclude from excavations, from bone and teeth finds, that the food situation dramatically worsened and that the incidence of disease (without professional medical care) increased. A Norwegian priest, who made a tour of inspection about 1350, reported that the *Western Settlement* was deserted; he found cattle wandering around, but no longer any human beings. Where had they disappeared to? Not even their corpses were discovered; permafrost had made the old Viking burial practices impossible. Whereas researchers once tended to believe in a theory of genetic degeneration, our present knowledge of fourteenth-century climatic deterioration suggests that harvest failures, famine and disease were to blame for the wiping out of the Greenlanders.[63] The last report from the main *Eastern Settlement* in southern Greenland, which reached Bergen in Norway around 1410, spoke of the burning of a sorcerer. This was a sign of acute crisis, as such acts are unknown there from any earlier period.[64]

The case of the Vikings, exceptionally well researched on the basis of historical texts and scientific studies, shows that climatic factors alone do not get us very far when we are dealing with human cultures. The climate did become colder in Greenland, but the Inuit expanding there from the north had no problem with it. Their culture rested on hunting and fishing, not on agriculture and livestock breeding. Only an economy based on cereal-growing and pasture-farming, which the Vikings exported from Europe as a legacy of the Neolithic revolution, was doomed to go under.[65] Unlike the Inuit – who already wore furs, as the mummified remains of a family from a capsized boat show[66] – the European Greenlanders stuck till the end to woollen clothing that was inappropriate in Arctic conditions. Their economy also worked to undermine the basis of their existence, since livestock overgrazed the poor soil and further eroded it. Like the settlers themselves, the cattle grew ever smaller and more prone to disease. Yet the Vikings obstinately clung to their farmer's way of life. Findings from refuse dumps exhibit very little consumption of fish or wild animals.[67]

There has been much debate about the reasons for this stubbornness. The cooling itself – icy seas, dwindling fish stocks – may have hindered a changeover to fishing, but it is more likely that there was a cultural cause. The church may have forbidden Christians to adopt the Inuit lifestyle, and the lack of trade may mean that it was regarded as unseemly to have any contact with the pagans. Maladjustment was the downfall of Erik the Red's descendants.[68]

The decline of Iceland and Norway

Failure to develop fishing as a compensation for agricultural decline was also apparent in Iceland. It is true that the Danish government denied it a fleet of its own, but lack of inventiveness among Norwegian farmers was ultimately the reason why deep-sea fishing did not catch on at that time. Since, on top of declining soil fertility, cod disappeared from the ever colder waters of the island's fjords, traditional forms of fishing offered no way out. In many parts of Iceland it was given up altogether.[69] The farmers had little or no investment capital. Numerous farmsteads that had prospered since the time of the original settlement had to be abandoned, especially in the north, where advancing glaciers and pack ice cut off the fertile valleys by sea for months at a time from the rest of the country.[70] The emptying of farms and villages, often attributed to fourteenth-century population decline, was a consequence of the climate change that inflicted casualties in a long process lasting well into modern times.[71]

From 1969 to 1982 an interdisciplinary project investigated the phenomenon of 'desertion' in the Nordic countries. It turned out that new settlement was common even in Denmark until the high Middle Ages, when the number of localities became constant around the year 1200. Throughout Scandinavia, a farming culture stabilized only in the favourable climate of the high Middle Ages. In Norway, with its scattered farmsteads, there were two catastrophic periods in each of which 40 per cent of farms were abandoned: one in the sixth century,

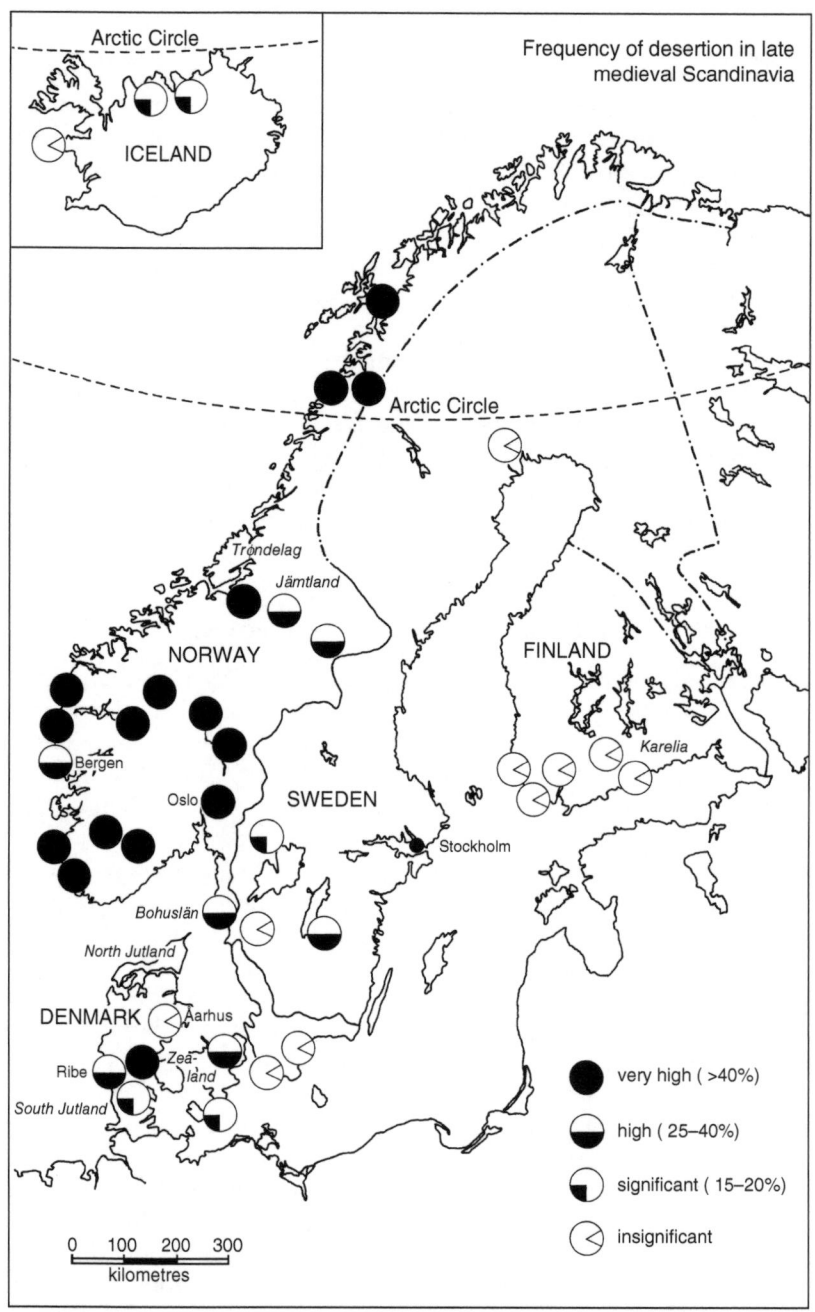

3.6 The worse climate of the Little Ice Age led to a changed settlement structure everywhere in northern Europe. We now have good statistics about the abandonment of farms and villages in Scandinavia.

during the pessimum of the *age of mass migrations*; and one after 1300 in the Little Ice Age. The desertion took place in a certain pattern. At more than 300 metres above sea level, the period of vegetation grew so short that cereal-growing became a risky business. When people in remote fjords could no longer be sure of feeding themselves, it became impossible for them to continue. The demographic consequences were considerable. From a peak around the year 1300, the population plummeted in the early fourteenth century, then declined more slowly before bottoming out around 1700.[72]

The advent of written administrative records provides us with detailed information about the impact of cooling from the seventeenth century on. Farmers could hope for tax forgiveness only if they had compelling reasons: a bad harvest alone was not enough. Thus, greater frequency of damage reports is a clear indication of conditions at a particular time. The rivers that flow into the Nordfjord in Norway burst their banks nearly every year between 1650 and 1750, and during the same period reports of damage due to glaciers and snow or rock avalanches kept coming in. In 1687 a particularly large number of farmsteads were destroyed by landslides. In 1693 or 1702 the flooding of pastureland was so serious that farmers could only pack up and leave. When the water retreated, the land was covered with sand, gravel and rock. Because of the danger from falling rocks, it became difficult for many farmers to find a single farmhand or labourer. It can be shown district by district how the number of farm animals (cows, sheep, goats) declined in the seventeenth century. In the valleys, most of the damage was due to flooding or to landslides after heavy rain or snowmelts.[73]

Wharram Percy and Britain's lost villages

The 'Deserted Medieval Village Research Group', which studies not only where people lived but also what happened to their farmland and cultivation patterns,[74] has pinpointed through aerial archaeology a large number of village desertions of which later intensive agriculture seems to have left no trace. Within just a few years it located more than four thousand such villages in England alone. First settled during the medieval warm period, they had flourished for centuries before they had to be abandoned after 1300. These are the 'lost villages of England'.[75]

Whereas similar studies in Germany usually blame everything on the plague, the better-researched English example shows that the causes were more complex and had to do with the climate change of the Little Ice Age. Population figures had already been declining since the great famine of 1315–22, before the arrival of the Black Death. Thus, Merton College, Oxford has shown that a windmill built at great cost in the 1290s no longer had a tenant in the 1330s, because there were no longer enough farmers to grind their corn there. In Upton (Gloucestershire) the Bishop of Worcester could no longer find any peasants to work his fields. He therefore converted them into sheep pasture – an interesting fact, because the decline in peasant numbers is often attributed to their eviction by capitalist sheep farmers. In the crisis of the fourteenth century the reverse was actually true: shortage of manpower and infertility of the soil led to the pursuit of sheep farming. Another type of conversion was from agricultural land to landscaped park, which began to take place in the fifteenth century.[76]

One of the best-researched deserted villages is Wharram Percy in North Yorkshire, which was gradually uncovered during the summer months of 1950 to 1990. This seasonal restriction on excavation work already answers the question as to why the village was abandoned. For, in the other months of the year, the climate in the uplands of the Yorkshire Wolds is unpleasantly cold and damp. Snow tends to remain on the ground in the valley where Wharram Percy is located, even when neighbouring ones are clear, such as that which contains its still populated sister village, Wharram le Street. The findings show that Wharram Percy was not simply abandoned in the early fourteenth century: thirty households were still living there in 1368, after the great plague had abated. Severe weather conditions led to further decline and eventual abandonment of the village. In 1458 there were sixteen households, but only one was still soldiering on in 1500. Wharram Percy was a wonderful place to live in the high Middle Ages; it was no longer that in the Little Ice Age.[77]

Anthropogenic changes

For a long time research into village desertion in Germany was under the influence of Wilhelm Abel's theory of agrarian crisis. In his view, the crisis was not due to such factors as a dwindling capacity of agriculture to feed people, but rather to the fact that population decline resulted in insufficient demand for food and falling incomes from land. Land prices began to fall in the fourteenth century, so that landlords no longer had an interest in occupying abandoned farms. The lesser nobility was impoverished, and robber barons contributed to the further decline of rural settlements.[78] This argument is no longer sustainable in the light of recent research into deserted villages in England and Scandinavia. For this has clarified the impact that the worsening climate of the Little Ice Age had on the environment and agriculture, and ultimately on the conditions of human life.

The abandonment of medieval villages, farms, pastures and fields partly occurred after population figures began to rise again. In the late sixteenth century, these probably stood at least at the same level as three hundred years before. As land for new settlement grew more scarce, more people had to be fed from much smaller areas. This was achieved through urbanization. Some cities, such as London, Paris, Milan, Naples or Istanbul, grew into metropolises and approached the quarter of a million mark. Others, such as Venice, Florence, Vienna or Amsterdam, climbed above one hundred thousand. But perhaps even more important was the fact that a large number of smaller towns – some four thousand in Europe – increased in size several times over. This created logistical problems for the supply of food, drinking water, building timber and fire logs, and for the removal of refuse and wastewater. Evidently there was a need for better cultivation and production methods and more ramified trade networks.

All this came at a price: the overexploitation of nature. More intensive industrialization and mining, increased energy needs for heating, smelting or salt production, supplies for large armies and the building of naval fleets again led to deforestation that did not pass unnoticed or uncriticized at the time.[79] Forest protection laws were enacted in many regions – for example, territories belonging to the Free City of Nuremberg or the Republic of Venice – but elsewhere there was no such precautionary environmental policy. In the British Isles, Spain or Italy, Dalmatia or Greece, Asia Minor or North Africa, scarcely any forest was left. This tree clearance had the well-known consequences: lowering of the water table (which especially in the Mediterranean intensified the problem of aridity) and greater susceptibility of cultivated land to soil erosion and flooding.[80]

Dance of Death

The Great Famine of 1315–1322

In the early part of the fourteenth century, Europe was visited by a disaster which those living at the time thought had never happened before: the Great Famine. It was a plague of a kind known only from the Book of Genesis (Gen. 41:30): the famous seven 'lean years'. The famine that struck Europe in the early fourteenth century lasted in many places for seven years, from 1315 to 1322. Early sixteenth-century chroniclers already called to mind the biblical plagues. And even modern historians have found no other event of its kind in Europe that lasted so long or had such a geographical extent. The Great Famine reached from the British Isles to Russia and from Scandinavia to the Mediterranean.[81]

Modern research has focused on four possible causes for the outbreak of this extraordinary famine: (1) the demographic pressure that built up during the high Middle Ages and outstripped the productive capacity of agriculture; (2) persistently bad harvest weather, which, given the inadequate food storage facilities, quickly led to the exhaustion of supplies; (3) food distribution problems in a context of wars and civil wars, which meant that regional harvest failure could result in a disastrous famine; and (4) the peasant conservatism that hindered adaptation to new environmental conditions.[82]

Chroniclers at the time thought the cause was clear enough: metaphysically, the famine was God's punishment on man, and materially it was the result of a series of natural disasters, chief of which was the abnormal weather. Long cold winters shortened the period of vegetation, while persistent rain damaged the crops, especially the grain on which 'our daily bread' depended. The French medievalist Pierre Alexandre has checked and compared the contemporaneous reports, and concluded that scarcely ever in European history has there been a series of cold winters like the ones between 1310 and 1330. At the same time, the second decade of the fourteenth century had the years with the most rain in the whole of the past millennium.[83]

From 1310 one cold wet summer followed another. The harvests, though bad, were at first enough for survival. But this would change in 1314, when in England and Germany heavy rainfall in summer was followed by a long cold winter, at the end of which the rivers burst their banks. The government of Emperor Ludwig IV 'the Bavarian' (r. 1314–47) could not have had the stars on its side: it had to contend not only with a rival king and the pope but also with the climate. The year 1315 was marked by persistently heavy rainfall, which began in mid-April in France, on 1 May in the Netherlands and at Whitsun in England; it lasted all summer throughout Central Europe. The sky was constantly overcast, the sun almost never to be seen, and temperatures were exceptionally cool. The Bad Windsheim chronicle reports that people began to eat dogs and horses, and uses the biblical metaphor of a Great Flood to describe the widespread flooding.[84]

The winter of 1315–16 was so cold that the Baltic Sea froze for weeks on end. The new year remained excessively cold and wet throughout, as floods destroyed mills or bridges and damaged industry or infrastructure. The Danube burst its banks three times in Bavaria and Austria, and on the River Mur alone (in Styria, Austria) high waters swept away fourteen bridges.[85] The harshest winter of the decade came in 1317–18, when the cold lasted from late November until Easter and it snowed in Cologne as late as 30 June.[86] Apart from such extreme events, the summer of 1318 was generally somewhat milder, but the years from 1319 to 1322 were as bad as the first three years of the cycle of disasters. There were terrible storms and floods in Normandy and Flanders on the North Sea coast, and excessive rainfall alternated with periods of drought on the continent.

Of course, war added to the general misfortune. After the magnificent reign of Philippe IV (r. 1285–1314), France went through the crisis period of the last three Capetians, Louis X (r. 1314–16), Philippe V (r. 1316–22) and Charles IV (r. 1322–8). Ludwig the Bavarian waged war against the rival king, Friedrich of Austria. But we know today that the scarcity of the time was not due only to wars. The general population suffered more than soldiers from hunger. Discontent fanned the flames of conflict. In 1315 the Swiss confederates fought successfully at Morgarten and won their independence from the ruling Habsburg dynasty. In 1314 the Scots wrested independence from England at the Battle of Bannockburn and extended the war to Ireland; a revolt broke out in Wales against English supremacy. The world seemed caught up in great upheavals. In Scandinavia the kingdoms of Norway, Denmark and Sweden became embroiled in dynastic struggles – everywhere wars, which historians explain at the level of *l'histoire événementielle*. Did they have a common cause in the shortage of resources during the Great Famine?

Already in 1315 a 'grim pestilence' was spreading in Europe, but it was not yet the Black Death. In Gelderland, in the Netherlands and in the Holy Roman Empire people spoke of the 'Great Death': a third of the population is said to have died in some areas. In the towns of England, France, the Netherlands, Scandinavia, the Holy Roman Empire and Poland, mortality was so high that new burial methods had to be introduced. Normally cemeteries were still within the city walls, but now they began to spread outside. In Metz, which cannot have had more than 20,000 inhabitants, 500,000 were said to have died – an exaggeration that gives some indication of the horror. Estimates refer to a death rate of 5 to 10 per cent in the year 1316 alone. And, although we have no reliable figures, one thing is certain: the Great Death was on the loose during the Great Famine of 1315–21. And with the Great Death came the Great Horror: reports arrived from England, the Baltic and Poland that parents were killing their children in their dire predicament, and that people were pouncing like cannibals on the dead.[87]

The 'Triumph of Death'

One of the icons of plague research is the Triumph of Death in Pisa's Camposanto, a large fresco, in the style of Tuscan painting, which recalls the prologue to Boccaccio's (1313–75) *Decameron*. In a delightful grove, a group of young people give themselves over to play and song. Then suddenly they are faced with the transience of life. Angels and devils fight on a cliff over the souls of the recently dead, whose corpses lie down below. The left side of the fresco shows the encounter of a merry hunting party with three dead figures laid out in coffins, from which snakes are already flickering their tongues. The images are flanked by scenes of the Last Judgement and the torments of Hell. This evocation of death on the mourners' route to the Pisan cemetery used to be seen as one of the most awe-inspiring reactions to the Black Death, an expression of horror at the mortality that could strike anyone down at any time.

Yet today we know from account books that this fresco was painted in 1338, nearly ten years before the outbreak of the Black Death in Italy. The dead do not bear the marks of the plague, nor is there any allusion to what are known in the plague literature as the concomitants of death. We also know today that several other pictures that used to be associated with the Black Death actually come from an earlier period – for example, the *memento mori* in the Dominican church in Bolzano, where Death on a galloping horse mows down the mass of the living. The confrontation with Death, the image of the Grim Reaper, the allegory of 'plague arrows': all this already existed.[88] One could shrug it off and argue that everybody has to die sometime. But it is not so simple. In the art of the founding father of Western painting, Giotto of Bondone (c. 1267–1337) – for example, in the Arena Chapel in Padua, which he painted in 1305 – we find little of the gloom of the next generation, to which the painter of the Pisan Camposanto, the mysterious Buffalmacco, belonged.

The solution to the puzzle is that in the 1330s, and again in the early 1340s, there were good reasons to be especially afraid of death. If we look at the climate history of Central Europe, we find after a few good years, which made people dream of a return to the warm period, a succession of difficult years in the mid-1330s. The summer of 1335 was cold and rainy; the wine turned sour and the harvest failed. The next year was also excessively wet. In 1338 people saw themselves facing biblical plagues: major flooding occurred in the spring; locusts arrived in summer and attacked Hungary, Austria, Bohemia and Germany, as deep inside as Thuringia and Hesse. They devoured large parts of the harvest. Only early snowfall put an end to the horror. But then many trees suffered damage from the snow, and the grapes that had survived the locusts were not untouched. The locusts returned in 1339 and 1340, until prolonged rainfall drove them away in August but led in turn to floods and crop damage.[89] The spring of 1341 was as cold as winter. The grain harvest of 1341–2 in England was so disastrous that tax relief had to be granted.[90]

The summer of 1342 brought one of the worst environmental disasters of the past thousand years. Intensive rainfall caused rivers to overflow in July, and the resulting floods destroyed the great bridges in Regensburg (Danube), Bamberg, Würzburg and Frankfurt (Main), Dresden (Elbe) and Erfurt (Gera). The high water carved deep gorges and permanently changed the landscape. The complete destruction of crops over wide areas led to higher prices and food emergencies. In 1343 there were again long periods of rain in July, August and September. Lake Constance burst its banks three times and flooded the towns of Lindau and Konstanz. The Rhine wrecked numerous bridges and buildings between Basel and Strasbourg. The harvest was spoiled by heavy rain and flooding. A cold wet spring, in which storms caused severe damage, delayed the tree blossoming season. In 1344 great aridity and drought reduced the size of the harvest, and only the wine came out of the heat well.[91] Famine in Italy cost thousands of lives in Florence, the great city of the Renaissance. The urban economy sank into a crisis heralded by the spectacular collapse of a number of firms.[92] This was the context in which the 'triumph of death' frescoes were commissioned – years before the great plague.

The Black Death of 1346–1352

The Black Death ranks among the greatest catastrophes in the history of Europe. Half the population is said to have died in just a few years.[93] This would mean that the effects were proportionately worse than those of the two world wars of the twentieth century together. Some historians ascribe to the Black Death decisive significance for the transformation of Western civilization.[94] More recently, however, a more nuanced view of the plague has emerged. Mortality during the first wave of the plague was probably lower,[95] and many religious practices ascribed to the Black Death years of 1346–52 either predated it (flagellation processions, representations of the dance of death) or fully developed only in the fifteenth and sixteenth centuries (cult of St Sebastian, cult of St Rocco).[96]

How was it possible for the Black Death to reap such an abundant harvest in Europe? The answer lies in the fact that the decades before its appearance had weakened the population and made it less resistant. Perhaps the Great Famine of 1315–22 played a role as the 'mother of all crises', since the stress of childhood hunger creates a greater lifelong susceptibility to disease.[97] The adverse climate of the 1330s affected the whole of the northern hemisphere. At that time there was great unrest among the Mongolian tribes, which advanced against China and contributed to the spread of the plague. Mass burials in western China are evidence that the great plague had its origins there and moved out along the Silk Road. In Europe the ground was already prepared: the winter of 1346 was exceptionally cold, with scarcely any warmer spells until June. It was already unusually cold on 22 September, and the still unripened grapes on the Main, Rhine and Mosel froze on the vine. The next year, 1347, was so rainy that the blossoming and harvest seasons were delayed: not even oats could be harvested, the year's wine was undrinkable, and snow was already falling in October.[98]

This was the situation in which the great plague reached Europe. The Tatars were the first group to be hit. In 1346, when they besieged the Genoese-controlled port of Caffa in Crimea, they catapulted plague corpses into the city – an early instance of biological warfare. From there the pestilence reached Italy in 1347 on board Genoese ships. In January it passed through Marseilles and then on to Avignon, seat of Pope Clement VI (1292–1352, r. 1342–52), whose personal physician Guy de Chauliac (d. 1368) wrote the first informed account of the disease. From Bordeaux it spread in June through several ports to England and northern France. It reached Paris in August 1348 and, in the same year, Bergen in Norway. The plague arrived in Germany either by land routes or through North Sea ports. The epidemic inflicted its heaviest casualties there in 1350, in Hamburg and other coastal towns, but southern Germany and Bohemia appear to have been relatively spared, as were parts of Sweden and Finland and the islands of Iceland and Greenland.

The victims were legion in Italian cities, however. Venice, where the disease began to rage in March 1348, lost more than half its population, Florence almost as many. The disease was completely unfamiliar to those affected by it, as the doctor Gentile da Foligno emphasized in his treatise. It was probably this novelty, together with previous weakening of the immune system, which helped the plague to be so deadly after many centuries of absence from Europe. According to the most recent estimates, 30 per cent of the population succumbed to it, with regional death rates between 10 per cent and 60 per cent. No other event in European history had such consequences.[99]

The main cycles of economic development

One of the great achievements of economic history is to have worked out long-term price series. These rest on a large number of local price series, such as those that Moritz John Elsas compiled in the 1920s from the account books of the Heilig Geist Hospitals.[100] We have no coherent price series for the period before 1200, since the transition from a subsistence to a market economy, or from an exchange to a money economy, was not yet complete and orderly accounts did not exist anywhere. The basis of a series is the price for cereals from which the 'daily bread' was baked. Bread was the most important staple in the late Middle Ages and the early modern period, so the evolution of its price influenced that of all other foods. Since the high Middle Ages there have been four major cycles, of which we shall here consider the first three as they are based on the same mechanism. In the course of these cycles, the price of bread rose over the long term, in a process that economic historians for some obscure reason call a 'price revolution'. The underlying basis for this was not a devaluation of money but a rising demand for bread. The population grew faster than the production of cereals. Agriculture reached its limits in the late sixteenth century – as it had at the end of the medieval warm period – and the poor could no longer feed themselves at affordable prices.

The English political economist Thomas Robert Malthus (1766–1834) described this situation in his famous *Essay on the Principles of Population* (1798): population displays a constant tendency to grow more strongly than the food supply; this leads to crisis, until mortality sharply increases as a result of disease and war, and the

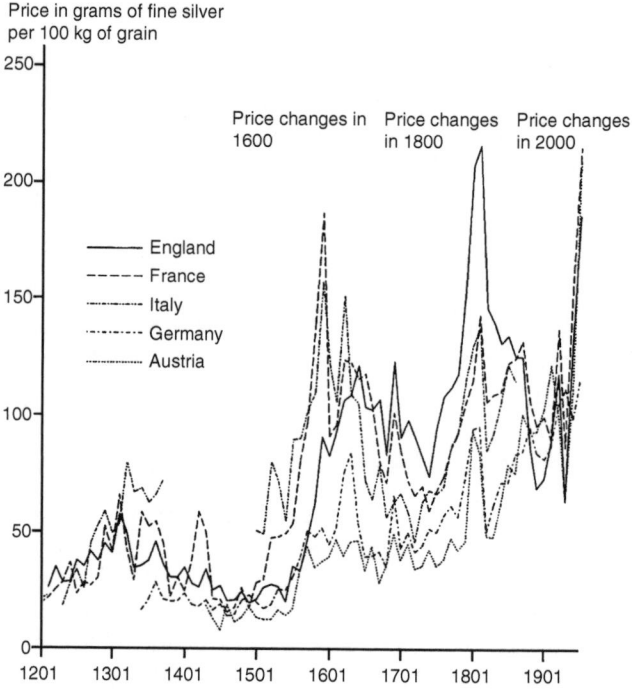

Price in grams of fine silver
per 100 kg of grain

3.7 The main bursts of social development: grain prices in Western Europe, 1201–1960.

total population is significantly reduced. The founders of demographic science called this the 'Malthusian crisis'. We can see from the long cycles of European grain prices that the crisis point was reached three times in preindustrial Europe: around the year 1300, in the second half of the sixteenth century, and around 1800. The precise moment of each crisis was determined, however, not by developments within society alone but also by the climate.

In the middle third of the sixteenth century – from 1530 to 1560 – a mild climate favoured population growth. By 1560 or thereabouts the total figure had probably again reached the level of 1300. Then the Little Ice Age entered a particularly unfavourable phase, with many cold winters, wet summers and poor harvests. Prices soared to previously unknown heights: in some months and years they were actually much higher than in the charts of economic historians, which flatten the peaks in mathematical averages stretching over a number

110

of years. Grain prices rose from the late-fifteenth until the mid-seventeenth century, to a level several times higher than the one from which they had started.[101] Since population growth then reached its limits, we speak of a *seventeenth-century crisis*. A whole series of wars, civil wars and revolutions then contributed to a fall in population, but in reality what we are talking about is a Malthusian crisis.

By then Europe had become a developed society. Shortages and price rises did not result in general impoverishment, but rather in rejections of the social and political equilibrium. In general, it was owners and dealers in bread cereals who profited from the high prices – for example, landowners east of the Elbe and sections of the feudal nobility in central and western Europe. The transformation of whole societies was linked to this process, as Immanuel Wallerstein in particular has emphasized in an application of dependency theory to European history.[102] Whereas a 'second serfdom' was introduced to raise grain production on the periphery of Europe, in Eastern Europe or the Spanish colonies, a merchant class rose to political dignity in the core of the 'European world economy'. Not by chance did the Netherlands experience its Golden Age at this time, when the rest of the continent suffered periodic famines. In all parts – including Italy or the Holy Roman Empire – there was a polarization between grain owners and the rest of the population. In northern Germany, where the nobility profited from the increased demand, people marvelled at the new wealth of the 'Weser Renaissance'. In Bavaria, on the other hand, laws had to be passed against the farmers, since they were selling their own grain and could sometimes afford more lavish weddings than their noble lords. The urban middle class did not participate in the grain boom: its income had less purchasing power in a context where wages held steady but product prices were stagnant and bread prices were sharply rising. This led to a serious deterioration in living conditions. Estimates based on shopping baskets at the time show that from the 1580s a healthy diet became difficult for a household of four and remained so for decades.[103]

Mortality crises of the Little Ice Age

Disastrous death-rates from disease were characteristic of the early modern period. In Augsburg, the city of the Imperial Diets in the sixteenth century, there were outbreaks of plague in 1519–21, 1533, 1543, 1562, 1572, 1586, 1592, 1602 and 1613. But these were completely overshadowed by the epidemics of 1628 and 1632–4, when nearly half the population of Augsburg succumbed. Since, at the same time, the Thirty Years War robbed the city of markets for its cloth, the Swabian capital never managed to recover and declined to the status of an ordinary small town.[104]

With a slight time difference, we find mortality crises in all European cities. In fact, the impression is unavoidable that, in the second half of the sixteenth century and the first half of the seventeenth, death-rates were higher than at any time before or since. In Italy the plague of 1575–7 assumed particularly grave proportions. The first cases of illness and death appeared in summer 1575, but as usual the epidemic receded during the winter months. It revived again in March of the following year and reached full force in the summer and autumn. In Milan, where the disease took 16,000 lives (one tenth of the population), the reform-minded archbishop Carlo Borremeo opened a special plague hospital. In Venice approximately a third of the 160,000 inhabitants died; public life came to a standstill, the schools were closed, gravediggers and layers-out had to wear a warning bell on their legs. This was the worst of the twenty-six outbreaks of the plague since 1348. When it finally subsided, the city thanked the Holy Mother of God by building the massive Santa Maria della Salute, under the direction of the famous Andrea Palladio (1508–80).[105]

Mortality crises were not caused only by the plague. Often they appeared in connection with other diseases, as a weakened constitution made people especially vulnerable. The typical ones in Europe were: louse-borne typhus, which drove people crazy with pain; smallpox, the pock disease especially dangerous to children; dysentery, the deadly diarrhoeal disease; and also measles, scarlet fever and influenza (which, because of its mutations, appeared in German under different names: 'English sweats', 'Spanish shivers', 'Bohemian sheep plague', 'catarrhus epidemicus', sometimes accompanied by coughs or whooping cough, and generally known until the Age of Enlightenment as 'the new ague'). Only with the pandemics of the eighteenth century and the founding of medical colleges and specialist journals was the disease properly diagnosed and given a scientific ('influenza') and a popular ('grippe') name. During the catastrophic mortality crises of 1632–5, famine took hold first as a result of war and harvest failure. Then came dysentery and typhus, then the plague.[106]

The link between famine and disease

Some of the literature disputes that a worsening diet in town and country led to greater mortality and susceptibility to disease.[107]

According to the Italian nutritional expert Massimo Livi-Bacci, scarcely one of the historical diseases was fuelled by malnutrition, and indeed food shortages actually impeded the growth of pathogenic organisms and decreased the incidence of disease.[108] Nevertheless, in the case of smallpox, a reinterpretation of English sources has now provided positive proof of increased mortality during the crisis years.[109] In the UN publications on famine in the twentieth century, infectious diseases such as tuberculosis,[110] typhus[111] and dysentery[112] are usually described as the result of undernourishment. And reduced fertility among women was statistically demonstrated long ago for the crisis periods of early modern Europe.[113] The declining intake of proteins (meat, milk, eggs) in the sixteenth century evidently had an effect: dental research and evidence of smaller body size from skeletal remains are indicators of malnutrition. People in the late-sixteenth and early-seventeenth centuries were on average shorter than ever before or since in the past two millennia, comparable only to those in the critical times of the early fourteenth century.[114] Contemporary chronicles underlined the link between famine and typhus: 'After this protracted rise in prices followed a terrible head disease, which, when it entered a house, cleared a large space and took off especially those who had barely a crust of bread to keep alive.'[115] In the famine crisis of 1570, for example, the number of publications on these diseases increased several times over.[116]

With regard to structural changes resulting from the worse climate, similar points may be made for all continents. Harvest failures led to increased mortality and susceptibility to disease, as well as to a decline in population levels. Even in China the demographic advance came to a halt in the seventeenth century.[117] We have statistical data for the Spanish Philippines and Dutch Ambon (now in Indonesia), as well as for Siam (Thailand), and in each case famines and epidemics were reported for the early seventeenth century. A sizeable percentage of the population in Java fell victim to the plague in 1625–6; a full third had died in the Spanish possessions by 1655. In the mid-1660s mortality rose sharply in Indonesia, as in Europe. Many of these phenomena have traditionally been explained away in terms of war or imperfect data, but the coincidence of events suggests a far-reaching economic or climatic basis. In so far as the affected areas already participated in international trade, the fall in export prices to Europe for such goods as pepper and other spices was another aggravating factor. On the whole, however, the crisis was due not to trade or the influence of colonial powers, but to problems of the local economy and society.[118]

Wartime violence and death penalties

In his *Triumph of Death*, which hangs in the Prado in Madrid, the Nederlandish painter Pieter Bruegel the Elder (1525–69) depicts a forest of gallows; it testifies to the frequency of the death penalty.[119] Early modern penal justice had to deal with ever larger numbers of offenders and responded by making legislation more stringent.[120] The treatment meted out was severe, and became ever more so in the course of the sixteenth century. Torture was used more widely than ever before or since,[121] the number of executions reached hitherto unknown levels all over Europe. The first thing that travellers saw as they approached the gates of large European cities around 1600 was robbers dangling from a noose. The 'Theatre of Horror' was supposed to frighten potential criminals and to convey the impression of an unshakeable social order.[122]

Much points to a general increase in violence, and the large number of victims in the Thirty Years War has often been especially mentioned. As much as two-thirds of the population is said to have died in this war.[123] On the other hand, we must remember that armies in those days were relatively small. The number of people killed in battle was actually quite low in comparison with those who died in the mortality crises due to disease. And, although the victims of marauding soldiers, 'strong-arm beggars' and other criminals suffered extreme violence in individual cases, they are not of great statistical significance.[124] The violence of war and death penalties were signs of an age in which resource shortages aggravated all possible (religious, social, political) conflicts. Yet, however bad the war and the violence, they are far outweighed by disease in local statistics of mortality.[125]

Special mention should be made here of China, which sank into an orgy of violence. The first half of the seventeenth century was a time of disasters, which may be compared in scale to those of the early medieval pessimum. In Yunan, in the mild southwest of China, snowstorms raged widely in 1601, and from then on periods of spectacular cold were reported everywhere in the empire. Although people died of cold, it was rather the aridity associated with cooling that brought Chinese agriculture to the point of collapse. Deadly famines plagued the country between 1618 and 1643. People lay dying by the side of the road. There was cannibalism, mass internal migration and various forms of violence; and in 1643 a great peasant insurrection under Li Cheng finally put an end to the Ming dynasty. Its violent overthrow cannot be thought of without irony, for it had come to power three centuries earlier in a similar climate-induced crisis that brought down the Mongol Yuan dynasty (1260–1368). The first Ming emperor, Tai Tsu (r. 1368–99), had been the leader of the insurgent peasantry. After the fall of the Mings, the Manchu dynasty came out on top in a series of bloody civil wars and managed to remain in power until 1912.[126]

Winter Blues

Psychological reactions to crisis

The Augsburg painter Barnabas Holzmann described as follows his reactions during the famine crisis of 1570: 'The bile from many sighs ran deep and bitter in my stomach, the sweet and sour in my mouth I could barely tell from each other. I have slept little, lain much awake, through many a long night . . .'[127] The American sociologist Pitirim Sorokin (1889–1968) once noted, against the background of the Soviet famine of the 1920s, that such catastrophes lead to extreme psychological reactions, to changes in thinking and feeling, in human conduct, social organization and cultural life.[128] There seems to be a consensus in the social sciences that social stress produces psychological disorders and a special kind of dejection.[129] The list of stress factors usually includes phenomena that appeared in a more intense form during the crisis years of the Little Ice Age: unexpected illness, death of a child or partner, family quarrels, loss of work or home, lack of sexual fulfilment, loneliness, involuntary childlessness due to infertility, physical violence, contact with crime, destruction of house or dwelling through fire, flood or other misfortune, financial crises.[130]

Seasonal affective disorder (SAD)

Among the many causes of sadness mentioned by the Anglican bishop Robert Burton (1577–1640) in his *Anatomy of Melancholy* are the endless days when dark clouds obscure the sunlight; some people suffer from these more than others.[131] This was true even in the British Isles, where people are not exactly spoiled for sunshine and the winter sun sets rather early in the afternoon. If one sinks into melancholy because of the winter gloom, there is only one remedy against it: to escape to Italy in September and to remain there for at least half the year. A recently identified psychological disorder is thus probably of special interest for the Little Ice Age: the famous 'winter blues'.[132]

The malady was well suited to an age of long winters and rainy summers, when in some years the sun was almost continually invisible behind dark clouds. Another new name for it, 'seasonal affective disorder', [133] indicates that the withdrawal of sunshine leads to exaggerated sadness and an increased risk of suicide. The acute symptoms include sleep disturbances, lethargy, eating disorders, depression, social problems, anxiety feelings, loss of libido and emotional volatility. Most of those affected complain of a weakened immune system and are more vulnerable to infections and other illnesses. Feelings of anxiety and depression disappear during the non-acute phase, but the symptoms of tiredness and disturbed sleep or eating remain in a milder form. [134]

If normal winters trigger such reactions in sensitive natures, how must things have been in the 'summerless years' for which the Little Ice Age was renowned? [135] We can only suppose that the kind of psychosomatic reactions familiar to us today also occurred then. Evidence of this is the 'tiredness' that gripped many people in northern Europe in 1784, when the sun grew dimmer following the eruption of the Icelandic volcano Laki. Light and darkness are culturally coded phenomena in European (and other) civilizations. Usually, light is associated with positive aspects and darkness with the force of evil. For this reason alone, increased darkness could not have made anyone happy, especially as there were limited possibilities for artificial lighting; pine chips and candles were poor ancestors of the daylight lamps now used to combat winter blues. To be sure, there were other reasons for mental depression in the Little Ice Age: for example, the harvest failures, livestock deaths and epidemic diseases – one smallpox outbreak carried away a third of the local population – that prolonged bad weather brought in its wake, as well as a stunting of plant growth due to acid rain. [136]

Despair and suicide

The climate of religious pressure and violence, recurrent agrarian crises, waves of beggars across the country, starving children with oedema in the streets, 'unnatural' diseases that doctors were unable to treat: all these phenomena, in addition to social tensions and wars, tended to foster psychological disorders. Mental depression was a source of great concern, as we can tell from the rash of consoling literature on sorrow, despair and melancholy. Already the second essay by Michel de Montaigne (1533–92), written in the 1570s, is entitled 'Of Sorrow', because 'generally the world, as a settled thing, is pleased to grace it with a particular esteem'.[137] Not only did the famine crisis spawn a literature of consolation; the sorrowful disposition was itself attacked as a 'melancholy devil'.[138] And Daniel Schaller wrote understandingly of 'the melancholy of men on earth': 'People lack nearly all courage, they are afraid and heavy-hearted, they look as a corpse, as a shadow, hang their heads down to the ground, as if they wished to crawl beneath the earth with their living body and would much rather be dead than alive.'[139]

Suicide figures reached previously unknown levels during the crisis years.[140] We should note in passing that suicide generated a number of interesting conflicts over the kind of funeral that should be held. Although no one else was physically hurt as a result, it was classified as a crime, literally as 'self-murder'. And murder was not a private affair, but an infringement of the divine order. The local population feared that the wrong kind of funeral might provoke God's wrath and cause further weather damage in a period already marked by harvest failures. 'To protect the harvest, livestock and human labouring power, those who kill themselves must be cast out from the community of the living and the dead.' As David Lederer has pointed out, the value system of the time reversed cause and effect. Climatic depression may have led to a greater frequency of suicide, but the popular notion was that 'suicide causes bad weather'.[141]

Melancholia was a fashionable malady of the age. Artists, intellectuals and princes discovered they had it – in France as well as Spain, where the respective dynasties of the Valois and the Habsburgs were both affected.[142] In England, where it reached a peak under Elizabeth I (1533–1603, r. 1558–1603), it used to be known as 'the Elizabethan malady'.[143] In a sign of the times, special treatment centres were built to deal with the mentally ill.[144] But the affliction had deeper roots in society: the autobiographical notes of the Puritan turner Nehemiah Wallington (1598–1658) show that the Protestant ethic, with its compulsion to prove oneself, may have plunged sufferers still deeper into their depression, since every external sign convinced them of the magnitude of their sins.[145]

The world under the melancholy planet

It is a pertinent symbol that the leading prince of the age, Holy Roman Emperor Rudolf II (1552–1612, r. 1576–1612), was considered to be melancholic, bewitched or insane.[146] Even well-intentioned figures such as the imperial ambassador in Spain, Count Hans Khevenhüller, reported the Emperor's peculiar disposition and

117

3.8 Reaction to an age of barrenness. Emperor Rudolf II has himself depicted by court painter Giuseppe Arcimboldo as the Roman fertility god, Vertumnus, c. 1591.

thought he had been born under 'the melancholy planet'.[147] Felix Stieve referred to pathogenic living conditions: 'his sexual dissipations stood in sharp contradiction to the religious views imprinted in him in youth; these still held sway over him, all the more because, in his fearful agitation, he must have been terrified by the idea of confession and responsibility before God'.[148]

The Emperor's bouts of depression may have stemmed from fears that we would still today regard as real – fear of the plague, for example, which several times forced him to escape to the countryside; fear of poisoning, so typical of the age; fear of political intrigues among his power-hungry brothers Ernst and Matthias, who did eventually depose him shortly before his death. Other fears we would rather define as irrational – fear of bewitchment (especially by Capucins or Jesuits), for example; or pangs of conscience over such sins as his passion for the daughter of his librarian, Jacopo Strada, with whom he indulged in excessive sexual activity and fathered several children out of wedlock.

Mentally disturbed princes could become a major political risk if their melancholy delayed decisions or paralysed the government, or if their lack of legitimate offspring led to a political crisis or, in extreme cases, to a war of succession. The childlessness of Henri III of France (1551–89, r. 1574–89) precipitated a new round in the wars of religion; that of Rudolf II lit the fuse to the Bohemian powder keg and led to the first military actions; and that of the melancholic Duke Johann Wilhelm of Jülich-Kleve (1553–1610) brought Europe to the brink of a general war, which was only fortuitously prevented by the murder of Henri IV (1553–1610, r. 1589–1610). The Jülich-Kleve war of succession spent itself in a regional confrontation, which was only later reinterpreted as a prelude to the Thirty Years War.[149]

Erik Midelfort concluded that we should seek the roots of the Thirty Years War not least in the madness of the rulers of the time.[150] But that was itself bound up with the psychological effects of the Little Ice Age. If witchcraft was *the* crime of the Little Ice Age, melancholy was its symptomatic illness. Some, it is true, were familiar with Aristotle's dictum that only the melancholic can achieve anything great in art or politics.[151] Yet the medicine of the age classified melancholia as a serious somatic disorder resulting from an imbalance of the bodily fluids (*humores*). According to the Galenian medicine that held sway in the universities, it was caused by an excess of black bile (melancholic humour) relative to blood (sanguinary humour), phlegm (phlegmatic humour) and yellow bile (choleric humour). The mixture of bodily fluids determined the temperament (or 'complexion') of individuals, their health, their range of activities and their view of the world.

Black bile was associated with unhealthy cold, melancholy with 'the windy, cold, dry season of autumn, the time of sad storms'.[152] An excess of black bile was thus to blame for fears, hallucinations, violent rages, apathy and deep sadness. And sadness made people – especially women, the weaker sex – susceptible to the devil's temptations. The Adversary promised wealth, sexual fulfilment and, if necessary, vengeance to the poor and lonely. The price was a pact with the devil, and the crime was called witchcraft. Johann Weyer (1515–88), physician to the melancholic dukes of Jülich-Kleve, saw a way to defuse the problem of witchcraft by defining it as a physical illness. He drew the radical conclusion that not only princes could be melancholic, but also the wrinkled old woman living round the corner: 'I have no doubt that, in her present cast of mind, the devil has confused her imagination with much and various mockery and delusion, duping and deceiving her so that she herself thinks these things have happened. But nor do I doubt that she never had any powers, and in this she is like other obsessed melancholics who are oppressed by the Fiend and believe that they are actually turned into a dog or a wolf.'[153] Clearly, existential anxieties and religious troubles were not only a preoccupation at the surface level of social conflicts; they also haunted people in their dreams.[154]

— 4 —

CULTURAL CONSEQUENCES OF THE LITTLE ICE AGE

The Wrathful God

On 28 December 1560, at a quarter to six in the morning, when the bells were pealing for early Mass, a light appeared in the heavens over Central Europe. At first it looked white, but then it took on a reddish hue and finally 'turned the colour of blood'. Those who were awake roused their neighbours, and lively discussions ensued. Many thought they saw in the north the reflection of a great fire, as if a whole country were in flames. At many places people began to sound the storm bells; danger seemed to be imminent. In Zurich the chief fireman rode out of the city, 'and in other places too there was a great glow; finally people followed one another back home and saw clearly that it was not something burning but a sign from God, meant to warn us all to lead better lives. Albrecht Küng, watchman at the Münster tower, informs me [. . .] that he has never seen such a sign in the sky as the aforesaid fiery and bloody sign. The Lord God mercifully wishes to grant us all his mercy, so that we may better and improve ourselves after this terrible and awe-inspiring sight, in his praise and honour and for our own good. Amen.'[1]

The Northern Lights, which appeared far south at the beginning of the early modern period, were interpreted – as were heavy snowfall, avalanches or flooding, but also harvest failure, price increases, disease and other effects – as signs from God, foretelling either the end of the world or divine retribution. 'Fiery signs in the sky are without doubt omens of the coming Day of Judgement, on which all elements shall melt from the heat and the world shall be cleansed by fire': this was the view of Johann Jakob Wick, for example, a collector of reports in Zurich.[2] The prospect of the world's end and divine

121

punishment could not really surprise people living at the time, or anyway the approaching disaster seemed easily explicable. 'A great accursed chain of events followed in the fiery sky; it lasted from January until mid-April. In summer terrible, awe-inspiring hail and windstorms ensued in this fiery sky, the like of which has never been heard or seen in human memory. The Lord God again wished to be merciful and to punish us only according to our deserts. A terrible pestilence followed thereafter and people died in Vienna in Austria, and in the following year of '62 in Nuremberg and other places besides.' The Massacre of Vassy, which inaugurated the civil war in France, fitted seamlessly into this apocalyptic scenario, which continued with harvest failures and epidemics.[3]

The subsequent boom in theological publications has traditionally been explained in terms of the growing religious tensions in the age

4.1 'On the great and terrible signs that have recently appeared in the sky and on earth. An epigram.' Title page of a pamphlet 1562. Since nature was seen as a sign system conveying messages from a wrathful God, unusual climatic events did not pass without comment.

of the Counter-Reformation. After the establishment of Lutheranism, the rise of Calvinism and post-Tridentine Catholicism brought two new, ideologically self-assured groups into conflict with each other. But only the rigours of the Little Ice Age can explain why their disputatious theological literature was able to meet the spiritual needs of the age. Their success becomes plausible once we realize that the insecurity of existence did not have only a religious dimension. Sermons and edifying literature often refer to natural phenomena and the disasters of everyday life; meteorological disturbances serve as a peg on which to hang theological considerations. Climatic events greatly concerned the population at the time. The literature of pastoral care developed a dynamic of its own, written for people in despair whose loved ones had just been torn from them, or who themselves quailed before chronic or acute illness and afflictions. The various denominations had to offer their own special prayers for all kinds of suffering. Manfred Jakubowski-Tiessen has argued that the rise of Good Friday to become the highest Lutheran festival was not the result of Luther's biblical exegesis. Rather, for the Lutheran church in the second half of the sixteenth century, it represented a new means of coming to terms with great pain and suffering, after the traditional bodily expressions (*mater dolorosa*, martyred saints) were no longer available.[4]

The ubiquitousness of death led to a new blossoming of the *Ars moriendi*. Authors usually situate the 'art of dying' amid the plagues of the Middle Ages, but in fact it reached a climax in the problematic decades around the year 1600.[5] The first book fair catalogues of Georg Willer (1514–93) already show the importance of the theme in the 1560s, when authors of all denominations began to publish about it in the vernacular.[6] 'Dance of death' imagery familiar from the late Middle Ages experienced a new boom, both in the literature of the day and through new editions of older works. Symbols of earthly transience – skeleton bones, skulls, death as a crossbow archer lying in ambush – were present everywhere in family registers, printed sermons, coins, copper engravings and paintings.[7] In the seventeenth century, *vanitas* symbolism finally achieved its greatest expression in the verse of Andreas Gryphius (1616–64).[8]

Actionism in the fight against sin

Quite a lot of people at the time felt that there was more misery, fighting and crime in society than ever before. That meant in the language of religion: more sins than ever before.[9] And human sin was seen as the cause of God's ire. Public attention therefore turned to moral failings as well as actual crime. Sexual offences, in particular, received sharply rising publicity, as the many kinds of premarital and extramarital sexuality seemed especially to provoke divine wrath. Serious crimes such as sexual commerce with the devil, sodomy, incest, bestiality and rape seemed to occur more often than in the past.[10] It may be suspected that only the 'constructed' crime became more frequent, and that increased attention to the whole complex of sin and crime rested upon sheer alarmism. Research has shown, however, that there was also a rise in violent and property-related crimes such as robbery and murder. If the records are to be believed, the number of thefts kept pace with the crisis years of the sixteenth and seventeenth centuries: not only the number of crimes but also the number of convicted criminals climbed to unprecedented levels. The rituals of deterrence with which the state reacted to this increase commanded widespread support among the population.[11]

The flood of legislation in the decades around 1600 answered a widespread need for regulation and order. In their response to the famine crisis of 1570, governments followed popular calls to punish the 'speculative buyers' [Fürkäufler] and 'usurers', who were blamed for high grain prices, and whose behaviour, understandable in terms of market economics, was regarded as not only 'selfish' but sinful and politically dangerous. Scarcely less common were accusations of blasphemy, a sin thought of as especially hateful to a personal God. The religious grounds for charges on this count were only superficially related to the growth of confessionalism; what they really involved was a supra-denominational 'economics' based on calculations of sin. Carnival or grain exports, witchcraft or usury, dancing or card games: these were all human sins that tarnished the honour of God.

In the eyes of Christian theologians, sexuality was closely bound up with sin, and its suppression in public life was a high-priority goal for early modern internal politics.[12] The imposition of moral temperance was a long-term objective, but it had scarcely ever been pursued as strenuously as it was in the second half of the sixteenth century, when attempts were even made in the countryside to introduce a strict regime of church discipline.[13] On the Catholic side, the ending of concubinage among priests and the enforcement of celibacy represented a major priority. Apart from adultery, another target was premarital sexual relations, which in many places – as in the famous custom of climbing a ladder to the beloved's bedroom window – was part of the preparations for marriage. Prostitution was another area of extramarital sex that was set to be stamped out. One town after another closed its brothels, which were often razed to the ground, as in the late example of the Free City of Cologne (1594).[14] Moral reforms were as rigid in Counter-Reformation Bavaria as in Calvinist Scotland.

This disposition to religiously inspired action was as dangerous in external as in internal politics. Contemporaries realized that witch hunts went hand in hand with campaigns for moral reform, which gave 'moral entrepreneurs' the ability to take the initiative. The 'crusading reformer', as the sociologist Howard S. Becker calls this type, 'operates with an absolute ethic; what he sees is truly and totally evil with no qualification. Any means is justified to do away with it. The crusader is fervent and righteous, often self-righteous. It is appropriate to think of [such] reformers as crusaders because they typically believe that their mission is a holy one.'[15] Such ideas were already viewed with mistrust: 'The witches of my neighbourhood,' wrote Montaigne, 'run the hazard of their lives upon the report of every new author who seeks to give body to their dreams.' And discussion with religious fanatics was no pleasure: 'I see very well that men get angry, and that I am forbidden to doubt upon pain of execrable injuries; a new way of persuading!'[16]

It is scarcely surprising that the crisis years led to social conflict. But it is less often appreciated that social conflicts in the broader sense of the term were expressed in conflicts over world-views and 'culture'. Social and religious unrest due to resource scarcity developed a dynamic of its own.[17] Harvest failures were the background to all manner of revolts and rebellions, including revolts among underlings, since excessive taxes and drudgery were especially oppressive in years of shortage.[18] Another reaction was scapegoating in the narrow sense, when violence was used against certain groups in the population or an abstract enemy image was constructed.

Anti-Jewish pogroms

Jewish minorities had been living since Roman antiquity in European provinces of the empire. When Christianity was adopted as the official religion, Jews became a tolerated religious community – a special status which, both in Byzantine lands and in the western empire, defined their position before the law. Since the Carolingian period they had been under imperial protection. And, despite Christian anti-Semitism, this construction also functioned under the early Capetians and the Ottonian and Salian emperors. With the First Crusade in 1096, however, imperial protection ceased to apply, and the 'crusader mob' got the idea that they should begin in their own country by liquidating enemies of the faith. The first massacres took place in cities with large Jewish communities, such as Rouen, Metz, Mainz, Worms, Cologne and Prague. Similar pogroms occurred at the beginning of the Second Crusade, in 1146 in France, and during the Third Crusade, in England in 1189–90.[19] The history of these anti-Jewish disturbances cannot be told here. But we should consider the extent to which Jews were scapegoated for the climatic rigours that set in with the end of the medieval warm period.

At the time when the climate was worsening, hostility to Jews entered a critical phase in Western Europe. The Dominican order stirred up accusations of host desecration and ritual murder. In 1290 the Jews were expelled from England 'for ever', only to be granted entry again after the civil war and revolution of the seventeenth century. Sixteen years later, Philippe IV of France (1268–1314, r. 1285–1314) followed the English example. In 1306 French Jews gave themselves up to the protection of the Holy Roman Emperor, Albrecht I of Austria (1255–1308, r. 1298–1308), and poured into French-speaking parts of the empire – Savoy, the Dauphiné and the Free County of Burgundy (Franche-Comté). In 1316 a new French king rescinded the expulsion order, but the Jews hesitated to return because, following excessive rainfall in 1314 and 1315, harvest failures had begun to result in price rises, mass hunger and death.[20]

After years of deprivation, groups of people in northern France marched off south in the hope of improving their living conditions. This movement of rustics and labourers linked up with Dominican friars, who stirred them up into what is known as the Crusade of the Pastoureaux or 'shepherds' crusade'. En route the Pastoureaux lived off plunder, and Jews were among those who bore the brunt. In many towns there were bloody riots. According to one Jewish source, 140 Jewish communities were destroyed in 1320–1 alone. The march gradually dispersed as it reached the Pyrenees, but the anti-Semitic fanaticism did not go away. In Aquitania rumours appeared of a Jewish plot in which lepers were poisoning the wells. The Jews were also accused of wanting to overthrow Christendom in league with the Muslim kings of Granada and Tunis and the 'Sultan of Babylon'. Here for the first time an external threat (Islam) was woven together with an alien religious group (Jews) and a deadly disease (leprosy) into a single complex.[21]

As far as we know, there was no epidemic at the time of the pogroms of 1320–1. During the Black Death pandemic, the accusations against the Jews changed: they were no longer agents but criminals, who had allegedly placed little bags of poison in the wells in order to bring Christendom to ruin. Fears of a plot assumed such dimensions that even Jews who had converted to Christianity were included in it. A search also began for Christian accomplices. There was rioting against beggars and lepers, as well as against monks. And the first systematic pogroms got under way. The starting point was the wave of killings in southern France (Toulon) in April 1348, from where they spread to the Dauphiné and Savoy, northern France, Spain, the Swiss confederation and German-speaking areas of the Holy Roman Empire, the Low Countries, Bohemia and Poland. It used to be thought that, after the outbreak of plague, flagellation processions went through the towns and stirred up pogroms. However, an exemplary reconstruction of events in several cities (Strassburg, Konstanz, Erfurt) has shown that often the exact opposite occurred: the citizens first murdered the Jews, in the hope of keeping the plague away. Then flagellation processions followed, and finally the plague arrived.[22]

Friedrich Battenberg noted that 'the favourable economic situation of the high Middle Ages until the end of the thirteenth century . . . allowed the comparatively peaceful development of Jewish communities and settlements'. But then the search for scapegoats began in the early fourteenth century, and pogroms criss-crossed Europe as the Black Death struck in the late 1340s. These spelled the end for urban Jewry.[23] Jews were blamed for many things – but it is clear that the weather was never one of them. No conceptual link was made between Jews and hailstorms, drought or extreme cold. We find in the literature a number of explanations for why pogroms became less common in the fifteenth century and eventually disappeared. The Jews in England and France were either killed or driven out. Spain followed suit in 1492, after the conquest of the last Islamic mini-state on the Iberian peninsula. Muslims and Jews had to convert to Christianity or else emigrate. Many Jews left by sea under the protection of the Ottoman empire, some heading for Italy, others for the Low Countries. In the territories of the Holy Roman Empire, Jews had become inconspicuous since resettling in rural areas. There were further expulsions, but these were less bloody than in the fourteenth century. The pogroms ended in Germany, although there were still numerous Jewish communities. The reason for this may be that Jews did not appear there as scapegoats for the Little Ice Age.

Witchcraft as the crime of the Little Ice Age

The English historian Norman Cohn (1915–2007) was the first to realize that in the fifteenth century witches assumed the role of scapegoat previously assigned to the Jews.[24] Witchcraft may be seen as the paradigmatic crime of the Little Ice Age, since witches were directly blamed for the weather, for infertile soil and infertile women, and evidently also for the 'unnatural' diseases that appeared in the wake of the crisis. As a social construct, the crime of witchcraft began its rise in the fourteenth century, running in parallel with the development of the Little Ice Age. Witch hunts reached their peak in Central Europe during the worst years of the Little Ice Age, in the decades before and after 1600. The crime disappeared from the penal catalogues after the end of the Little Ice Age, or with the emergence of more enlightened models of explanation.[25]

So much has been written in recent years about the European witch hunts that one may refer to a specialist literature for the details.[26] The alleged witchcraft was not simple magic but involved the image of a satanical sect, whose terrible deeds could not be credibly pinned on the Jews. This new crime arose in the very region that had initiated the anti-Jewish pogroms following the Great Famine and the Great Plague. In fact, some of the old terms were now applied to the new crime: witches met on 'the Sabbath', the Jewish holy day, and their gathering was known in Savoy, the Dauphiné and western Switzerland as a 'synagogue'. The two became condensed into the 'witches' sabbath' – a term which has remained in existence, without calling to mind the anti-Jewish persecutions of the plague era. At first regional terms were used for the new crime; one of these, the southwestern Swiss *hexereye* [today's *Hexerei*], was then taken up and employed in large areas of the German-speaking world. Apart from curses, *Hexerei* included some quite incredible things such as pacts with the devil, coitus with the devil, flying through the air to the witches' sabbath, and a capacity for human to animal metamorphosis.[27]

Witches were completely different from female 'healers' or sorceresses. They were imputed with powers that popular superstition had ascribed to fairies: they could sneak into houses through cracks or keyholes, enter cellars and drink wine-casks to the bottom, without any perceptible loss of wine. They could gobble up animals and then put them together again in such a way that no one saw a difference. They could make people fall in love, or find and restore things that had been lost or stolen – but in the *Interpretatio christiana* this was

128

4.2 Anthropogenic climate change. European superstitions repeatedly blamed humans for climatic extremes. This woodcut from 1486 shows a sorceress conjuring up a hailstorm.

all a fraud, because they were in league with the Evil One and bound to him by ties of the flesh. Their primary task was to cause mischief. In reality they contaminated wine and animals, sowed disease and discord, killed and ate children – here we find an echo of the charge of ritual murder – and called down impotence on men and infertility on women, animals and crops.

This accounted for some of the main problems that people faced in the Little Ice Age: childlessness, livestock epidemics, repeated harvest failures, often mysterious diseases. Cows that gave too little milk, sudden death of children, late frosts, persistent rain or sudden hailstorms in summer: the search was on for someone to blame for such devilry. The idea that misfortunes of this kind happened by chance was alien to many Europeans of that period. Witches were the scapegoats that people needed to explain the disasters of the age.[28]

The problem with witches was that they practised their crimes in secret and left scarcely any evidence of their guilt. Therefore torture played a central role in witch trials. However, unlike among bandits or pirates, torture could not simply be applied at random in European countries. According to Roman law, which was taught at the universities, torture was permitted only under certain conditions: there had to be a definite suspicion of guilt, circumstantial evidence and concurrent statements about the reputation of the accused. In some countries where Roman law had not been adopted – England, for example – torture was not permissible. In either case, confessions of witchcraft could be obtained only if the provisions of the law were disregarded and physical pressure was brought to bear in an illicit way, or if women confessed of their own accord.

The age of witch persecutions

If we take into account the growing numbers of people in Europe who were opposed to the persecution of witches, as well as the considerable legal obstacles to the killing of individuals because they were thought obnoxious, it will come as no surprise that far fewer were burned as witches than is often supposed. Figures of nine million or more appear in some of the literature, but the contemporary size of the population alone makes that impossible to accept. In fact the total number of victims was probably around fifty thousand.[29] Witch trials were not generally accepted in the fifteenth century – which meant that those who advocated them had to face stern rebuffs. The author of the *Hexenhammer*, the inquisitor Heinrich Kramer, was expelled from the Tyrol because of his illegal methods.[30] In northern Italy, after persecutions in the diocese of Como, a violent debate broke out in which Franciscans and legal experts took a stand on the unlawful practices. Few witches were put to death in the decades between 1520 and 1560. This used to be explained in terms of the laudable influence of the Reformation. But the disturbing fact remains that Martin Luther, Ulrich Zwingli and John Calvin all advocated the execution of witches, and that statutes permitting it were adopted at an early date in their respective capitals, Wittenberg, Zurich and Geneva. In the light of climate history, it should be pointed out that in the 1520s the Reformation and the Peasants' War raised other problems that weighed on people's minds, while the particularly mild climate after 1530 did not lend itself to too many accusations of witchcraft. Only after 1560 did the subject arouse wider interest: those were the worst years of the Little Ice Age, and every kind of disaster that could be laid at the door of witches became more acute. The decades between 1560 and 1660 were *the age of the witch persecutions*.

4.3 Media depiction of the Little Ice Age at its height: 'On the noisome and dreadful hail that occurred this 6 July 1561, and on other storms, tempest, etc.' A drawing in Johann Jakob Wick's collection of news reports.

In fact, the witch persecutions began after the disastrous cooling in 1561, the summer storms of 1562 and the subsequent harvest failures and widespread epidemics. At the same time, an interesting debate flared up over the causes. The man who started it, the Protestant preacher Thomas Naogeorgus, blamed witches for the bad weather, harvest failures and epidemics, in sermons that were printed soon after delivery. Two Württemberg deans objected that not witches but God alone was responsible for the climate; this was strictly in line with the arguments of the Reformer Jacob Brenz (1499–1570), who nevertheless concluded that witches should be executed. Johann Weyer then took up the cudgels, accusing Protestants of inhumanity because they wanted to execute people for a crime they could not in fact have committed. In his monumental work against the persecution of witches, published in 1563, Weyer argued that the crime of witchcraft simply did not exist but was a devilish illusion.[31]

As the great persecutions got under way, the main arguments against the scapegoating were also presented – and the relatively low number of executions shows that many cities and states accepted them. It has even been suggested that large countries with intact institutions – England, France or (in the Holy Roman Empire) Bavaria, Kur-Brandenburg, Austria or Saxony – did not permit large-scale persecutions because there was always a body that blocked such efforts. The institutions in question would have been the church, the state administration, provincial governments, universities, city magistrates, corporate organizations, royal courts, nobilities and, in Catholic countries, large monasteries or religious orders. The impetus for the witch persecutions came neither from the church nor from the state; it came 'from below'.

Precisely for this reason there is a strong link between the Little Ice Age and witch persecutions. After the cold wave in the early 1560s, the hunger crisis in 1570 or the years of harvest failures in the 1580s, a pogrom mood prevailed in many areas. The peasants wanted to 'seize the evil by the roots' and to 'clear away' the guilty ones. Chiliastic expectations attached themselves to witch-burning: the crops and the wine would supposedly improve afterwards, and the children and cattle would no longer die of mysterious diseases. The chronicle of the Trier canon Johann Linden put it quite dramatically: after years of grain harvest failure under Prince Bishop Johann VII of Schönenberg (r. 1581–99) – only two out of seventeen were fruitful – 'the whole land rose up for them to be wiped out'. When the sun finally shone again and nature apparently returned to normal, many thought that this was due to the witch persecutions.

But the cold years came back later in the century and early in the next one, and the Little Ice Age headed into a particularly inauspicious phase. This was when the European witch hunts reached their peak, with the main centres in Franconia and the Rhineland. After severe frost in late May 1626 ruined the fruit harvest and froze the vines, thousands of people were burned as witches in the following years in Bamberg, Würzburg and Aschaffenburg – not only women from the lower classes, but city councillors and their families, sitting mayors and even an occasional nobleman or theologian. This could happen only because criminal procedure law was set aside, with the argument that an exceptional crime called for exceptional measures. Arrest, torture and conviction without sufficient evidence were the order of the day in territories with deficient institutions. In small states such as the German prince-bishoprics, which lacked major towns, a functioning parliament, a strong nobility or an experienced law faculty, there was no resistance to an unlawful procedure that was already criticized by people at the time and is today regarded as a symbol of injustice.[32]

Sin Economics as the Motor of Change

In full accordance with the Old Testament, early modern theologians of every denomination interpreted climatic extremes, hailstorms, floods, harvest failures, pestilence, shortages and hunger crises as God's punishment for man's sins. Every kind of disaster, including the evil deeds of witches, was the result of human transgression.[33] We shall therefore speak here of sin calculation or a 'sin economics', because an attempt was made to convert the classical argument of divine retribution into a system of constantly updated calculation. The greater the sins, the greater the punishment. Whether in Munich or Zurich, Dresden or Geneva, not only lay people but also theologians such as Heinrich Bullinger (head of the Reformed Church in Zurich) implied that there was a kind of collective 'sin account', which could be overdrawn only on pain of penalties. And the penalties affected not only the individual but whole groups or the whole society.[34]

The sin economics of the time produced the key link between nature and culture: it was the mechanism that helped a meteorological event to acquire its social significance. Preachers such as the Bamberg auxiliary bishop Jacob Feucht (1540–80) with his 'five sermons on the great dearth, famine and violent storms',[35] the Zurich Reform theologian Ludwig Lavater (1527–85) with his *Sermones* on dearth and famine,[36] or the Lutheran theologian Thomas Rörer (1542 to post-1580)[37] posed the justifiable question, 'why in countries, towns and villages everything is increasingly poor and debased'.[38] They were operating with the same stock descriptions of God's anger at man's sins that we can find in the Bible.[39] Unlike the sins against the environment of our own age, which are blamed for nature's punishment or even 'revenge', theologians of the sixteenth and seventeenth centuries expected punitive action on the part of a personal God. And such actions might relate to all possible crimes and spiritual or moral transgressions.

The sin economics that theologians used as their main tool to interpret the worsening climate did not normally require a scapegoat. Rather, moral apostles of the time sought to change how the great mass of the faithful behaved, urging them to sweep before their own door instead of shifting their guilt on to minorities. As all the major Christian churches shared this approach, with the support of the respective states whatever their political constitution, sin economics played an exceptional role in the cultural changes that occurred

4.4 Typical scapegoating in the Little Ice Age. A mass burning of alleged witches on 10 June 1587 by the princes of Waldburg-Zeil. But people already disputed the purpose of this judicial killing, as the climate rarely improved, however many witches were burnt.

during the Little Ice Age. Here, however, we need to start by making a distinction. Sometimes changes were truly rational only at a technical level and were directly associated with climate change; or sometimes it is difficult to separate the moral from the rational justification.

Consequences of cooling for everyday organization

There were obvious reactions in many areas: protective measures against rain, cold or snow; changes in clothing, architecture, heating or wood management. Longer heating periods were not only a cost factor but had implications for the environment. The need for wood increased and led to shortages or resource conflicts.[40] Diary entries make it clear that transport added to the cost of the annual purchase of wood, and each delivery required labour, logistical skill and storage space.[41] Besides, the wood itself took longer to grow in the colder climate, as we know from dendrochronological evidence. Or, as Schaller put it: 'The wood in the forest does not grow as in former times [. . .]. It is a common complaint that, if the world were to last much longer, it would soon eventually run short of wood and break down.'[42] There were also architectural implications, as large buildings had many more rooms in need of heating than in the Middle Ages. The greater space in castles was a sign of growing wealth, but it also meant that a single set of heated apartments was no longer enough. Heating became necessary to survival, and this made the man in charge of it more important in castle life. In Hradschin, the Prague castle (now the Czech Hradčany), only the director of heating had the keys to all rooms; he was the first to enter them each morning to make them habitable.

Changes in the cityscape as 'Gothic' houses gave way to 'Baroque' at the end of the sixteenth century may also have been related to the climate. Admittedly these were long-term trends, but stone did increasingly replace timber in housing construction around the turn of the century. This permitted a more rational use of fuel and contained the danger of fire despite the longer heating periods. The tendency to increase the size of houses, both vertically and horizontally, reflected not only the greater prosperity of many groups in society but also the need for larger stocks in a time of recurrent supply crises. This was true of private individuals as well as public authorities – when they could afford it, that is. Servants and day labourers, who could not afford a place of their own, often lived in the upper storeys, which benefited from the heating lower down. More living space was also necessary because the moral conceptions then in the ascendancy demanded a gender separation of domestic servants; the children of the house were also more rigorously kept apart from service personnel, and fewer people than before shared a bed within the family. This separation of beds and bedrooms helped to limit the communication of body vermin and disease. The spatial barrier between humans and animals also reduced the danger of epidemics.[43]

The spread of glass windows in the sixteenth century – instead of the old shutters, paper, turpentine-soaked canvas or animal skin – improved the conservation of energy. Open hearths made way in bourgeois homes for luxurious tiled stoves, which saved energy, restricted the circulation of smoke and therefore made domestic life generally more comfortable. Stoves became display objects, decorated with tiles and hot plates.[44] The thick feather beds and cushions often described by late sixteenth-century travellers to Germany were a response to the colder conditions.[45] Four-poster beds, now to be found in bourgeois houses too, protected people against cold as well as vermin. And we may suppose that the same was true of the popular wood panelling. Wooden floorboards were cheaper than stone and provided better insulation. During the long winters, hypothermia and related colds and illnesses were a real danger, which had to be fought against with suitable furnishings and clothing. The new woollen nightdress, which Weinsberg had made for himself in September 1582, came down to the feet and was stuffed with fox fur.[46]

The case of clothing illustrates how quickly the issues become complicated. It was clear to people living in the second half of the sixteenth century that style had been undergoing fundamental change. Weinsberg devoted a whole chapter of his *Denkwürdigkeiten* to the 'manifold changes in dress' since the days of his youth. Traces of the cold climate are everywhere to be seen in these 'reminiscences'.[47] In 1570–1 he reported that since Christmas 'more snow has fallen than I have seen in my life, up to my knees and waist. And in a number of streets of Cologne it was impossible to travel by coach or cart before Lent. In a number of streets the snow was piled up like a dike, so that you couldn't see across to the other side.'[48] He noted, in the climatically favourable Rhine valley, that 'the air is so cold and biting that you can clearly see and recognize the breath from your mouth'. Given that the heating was often inadequate, the cold forced changes in people's behaviour. Meals were eaten in a special room when it became too cold in the normal winter dining-room. In bed Weinsberg took to wearing a woollen, fur-lined sleeping costume and a soft knitted nightcap, and more generally he protected himself against the wet conditions and the cold storm winds.[49]

Sin economics and clothing

The sartorial changes went beyond what one might expect as a direct response to the colder climate. All the evidence suggests that the clothing of the upper classes was lighter around the year 1500 than it became later. In court paintings of the early sixteenth century – for example, the famous portrait of King François I of France by Jean Clouet (Louvre, Paris) – the clothing emphasizes the body through its ingenious composition and sophisticated cut; the richly flowing material is evidently a lightweight silk. The men's tight leggings and the women's plunging neck-line – which can be seen in drawings by Dürer and Holbein – would scarcely have met with the approval of Reformation preachers, but they were widespread nonetheless. The playful use of clothing material – for example, multicoloured pantaloons with slits that show glimpses of coloured underwear or bare skin – aroused the anger of envious theologians, as in the famous *Vom hosenteufel* by Andreas Musculus (1514–81), which raged against the 'indecent devil's pants'.[50] Female dress offered erotic excitement with its 'holed sleeves' and 'little golden chains'.[51] Not for nothing were there devotional references to 'frock devils', 'pantaloon devils', 'frill devils' and the like.[52] Fashion reflected the climate of the time, but also more than that: the spirit of an age whose *joie de vivre* had not yet been deflated by religious gravity.[53]

Fashion histories record a switch to widespread use of heavier materials in the course of the sixteenth century. In contrast to their quite risqué appearance early in the century, the upper classes of every nation and denomination, age group and sex, are later seen wearing high necks, their bodies hidden beneath dark heavy materials. At more or less the same time, undergarments in both the narrow and the broader sense began their triumphal march: those worn directly on the body, beneath layers of shirts and doublets, such as Weinsberg had made for himself in September 1585 and had to wear again in 1586 until late May because of the great cold;[54] but also articles of underclothing (bodices, dress frames, wire supports) that gave the visible clothing its desired shape. Body movements became more 'dignified' among the upper classes, and even the shape of their bodies may have been affected. Protection against the cold – which was probably what led Habsburg crown princes and the English Lord Chancellor Francis Bacon to wear those grotesque tall black-felt hats,[55] or Weinsberg to put on warm slippers at home and to buy a special pair of shoes to wear 'in the street when it is icy'[56] – was only one partial aspect of the change.

The 'Spanish style' was more than just a fashion. It took the raised moral demands into account and, in addition, reflected the colder weather. As we can see in the dual portrait that Hans von Aachen (1552–1615), imperial court painter, produced of Duke Wilhelm V ('the Pious') of Bavaria and his wife Renata of Lorraine,[57] the change ran literally deeper. Bodies were hidden beneath massive black material invisible in the dark, and heads separated off by high stand-up collars. The female figure was caricatured in an almost triangular form, with ruffs that must have made any movement difficult and compelled the stiff posture visible in pictures from the time. The floor-length dresses were supported with iron or fishbone rails; legs and neck were covered, and hands often wrapped in gloves. Like the men, women wore dark heavy headgear, plus furs and muffs in cold weather.[58] The stiff overlarge black-felt hats, the plate-sized white ruffs, the wide black cloaks, the heavy boots and the gloves were worn at the various European courts,[59] in the Holy Roman Empire, in the Netherlands, in England, and even in Spain. Hats were worn not only in the open air but also indoors.[60] This had an effect on the hair: Weinsberg wrote that in his youth people had worn it over their shoulders, but that now they kept it short.[61]

In the culture of popular festivals, it was not easy to get killjoy restrictions accepted – as we may gauge from the moral writs that imperiously denounced the wearing of décolletés, short dresses or tight breeches. The issue in these moral-sartorial writs was not the cold but human sinfulness. Women's clothing excited the imagination of moral reformers to a quite special degree. Another bone of contention was 'the use of make-up, the painting and smearing of foreheads with all manner of strange colours'.[62] On this the fingers were raised high in admonition: 'Because women, for their appearance, use quicksilver, snake grease, adder, mouse, dog or wolf dung and many other disgraceful and stinking things that I for shame's sake may not mention by name, rubbing and anointing their forehead, eyes, cheeks and lips with poison, they may for a short while have a glowing and shining countenance, but before long they become all the more disgraceful, scruffy, appalling, shapeless and old, and in their forties look as if they are seventy.'[63]

Painting in the Little Ice Age

One of the most moving paintings of the early modern period was commissioned by Emperor Rudolf II at a time when he enjoyed better mental health. The earthly head of Christendom had himself portrayed by his court painter Giuseppe Arcimboldo (1527–93) as a pagan god of fertility, his face composed of exotic and indigenous fruits, ears of corn, sweet chestnuts, maize, courgettes and peaches – an arrangement with distinctly surrealist features.[64] Infertility was a problem of the years around 1570, when the most terrible hunger stalked central Europe. The proposal of Arnold Hauser (1892–1978) to view the Mannerist style as an indication of growing social instability and insecurity in a shattered age,[65] and the Baroque as the recovery of a new equilibrium, has met with little enthusiasm among art historians.[66] The same is true of Hans Neuberger's attempt to prove climate change by counting the number of clouds in landscape paintings;[67] after all, few clouds are to be expected in medieval canvases with a golden background, or in twentieth-century abstract art for that matter. On the other hand, it is amazing how often dark clouds appear in the work of such artists as El Greco (1541–1614). It may be that the shape of the sky in his *Disrobement of Christ* contributes to the 'spiritualization of the theme',[68] but there is nothing to say that it should not be related to actual changes in nature. Sombre clouds even appear in the background in wood cuttings and copper engravings, although they are hard to represent with such techniques. Apart from genre paintings of the seasons, the weather first features in large canvases – for example, *The Dark Day* by Pieter Bruegel the Elder (1525–69), where 'under a grey sky, in which a cold wind chases grey clouds in front of a pale moon, . . . the steel-grey mountains with their snowy peaks' rise over a town that looks frozen solid; a storm puts boats in distress on the lake and wreaks havoc in a village close to the shore.[69]

Even more suggestive is the invention of a new branch of landscape painting: the winter landscape. Bruegel's *Return of the Hunters*, the prototype of this genre, radiates a large measure of harshness and hopelessness in its tired colours.[70] The icy bluish-grey sky transcends the genre features of a seasonal portrayal. 'In *Return of the Hunters*, each shape participates in its way in the essence of winter: the trees appear more driven into the earth than grown out of it; the boughs are dried and brittle in their complex entanglement; the houses seem to shiver as they cower beneath their hats of snow; the pointed shapes of the mountains make them look like blocks of ice; men and animals

139

4.5   The extreme climate of the 1560s led artists to invent the winter landscape. Pieter Bruegel the Elder, *Return of the Hunters*, c. 1565.

turn into grim silhouettes, akin to their own shadows.'[71] Bruegel probed many variants of the winterscape theme in the last years of his life: for example, in *Census at Bethlehem* (Brussels, Musées Royaux des Beaux-Arts) from 1566, or *Child Murder in Bethlehem* (Vienna, Kunsthistorisches Museum), in which the Roman soldiery rampage in a snow-covered Dutch village. Their leader bears the features of the Spanish governor of the Netherlands, Fernando Álvarez de Toledo, Duke of Alba (1507–82): Roman and Spanish terror, biblical imagery and national history pass seamlessly into one another. One of Bruegel's most startling examples, *Adoration of the Three Kings in the Snow* (Winterthur, Sammlung Reinhart), has nothing charming in it. Indeed, it is difficult to make out the Bethlehem crib in the grim winter landscape with heavy snowfall.[72]

Iconographic representations of other disasters may be ranked among the winter landscapes: for example, floods, collapsed mountainsides, storm tides, distressed boats and shipwrecks. The last of these, in particular, is fraught with such symbolism that scarcely anyone has looked for a real-life incident behind it.[73] Yet, in an age of frequent storms, shipwreck constituted a very real danger, not only for crew and passengers but also for shipowners and those who provided the capital for trading expeditions.[74] Lives and property were at stake, and the newly developing insurance business could do no more than reduce the risk. In February 1576 Khevenhüller reported the loss of eight Spanish galleons with crew and money shipments on board, in a violent storm in the harbour of Villafranca.[75] In the case of landslides, avalanches and flooding, people assumed a direct connection with the quantity of rain and did not look for an allegorical significance. It was the allegory of collapse that made tempest and shipwreck so attractive to artists.

'Worry, sorrow and fear': Global cooling in music and literature

The influence of the Little Ice Age on music is too complex for us to do more than mention it here. The fact that, in one dramatic hunger crisis, the great Orlando di Lasso (1532–94) switched genres and published 'repentance psalms' can hardly be explained only by the threatened closure of the Munich court chapel, then in dire financial straits.[76] The composition of such psalms was not without a certain logic, in a time of disasters when the authorities were introducing additional prayers and days of penance to placate a wrathful God. The turn to penitence and new kinds of musical setting was an international phenomenon, noticeable among leading composers in Spain and Italy such as Giovanni Pierluigi da Palestrina (1525–94). The task of coming to terms with disease and sudden death began to assume a disproportionate role in the Lutheran body of song.[77] Music historians will have to discuss the extent to which formal trends such as the employment of basso continuo, the suppression of the polyphonic tradition, the fondness for dissonance or the introduction of bar lines may be interpreted in the light of changed mentalities.[78] The emergence of major new forms such as opera may have been linked to political-structural change and the increased role of the court, and only in this way, against a background of climate change, can they be seen as part of a centralized strategy to come to terms with the crises.[79]

In literary history, what stands out most is the rise of miraculous and spine-chilling literature, which concentrated on signs from God in miscarriages, miracle births and satanic births, on unknown animals and plants, and on monsters and misshapen forms in nature.[80] This predilection had a counterpart in handicrafts and both academic

4.6 Excessive precipitation resulted in landslides. The Swiss town of Plurs in the canton of Graubünden, before and after its destruction in 1618.

and court art, where painters, goldsmiths and sculptors looked to create unusual forms – the famous contorted shapes of Mannerism, which found a continuation in the irregular pearls of the Baroque.[81] The search for the peculiar, for portents in nature, had a counterpart in the royal collection of unusual objects, the 'cabinet of curiosities', an embryonic systematization that tried to bring the unfamiliar under a new order.[82] The comets of 1577 and 1618 unleashed a flood of publications that became significant not least for political history and the scientific revolution. But other signs in the sky were also recorded with amazement and dread: for example, the previously mentioned Northern Lights, 'blood rain' (coloured deep red), 'wheat rain' (containing grains of wheat), or the ostensible sighting of 'fiery dragons'. Together with reports of crimes, executions, devils and witches, these formed the basic material for the printed broadsheets that became typical of popular literature towards the end of the sixteenth century.

One new genre on the rise after 1560 was a special kind of Protestant edification literature, *Teufelsliteratur*, which attached a particular devil to certain human failings. The *Saufteufel* castigated alcoholism in all its forms, the *Hosenteufel* crazy fashions of the time, the *Spielteufel* (1561) insufficient attention to the Beyond, the *Fluchteufel* (1561) the sin of blasphemy, and Ludwig Milichius's *Zauberteufel* the supposedly greater preoccupation with magic and rampant superstition. This type of moral criticism found its first summation in 1569, in the *Theatrum Diabolorum*;[83] shortly afterwards a second 'theatre of demons' book appeared with twenty-four new texts, and a further sixteen were written and published before the turn of the century.[84]

After the beginning of the witch persecutions, a *Theatrum de Veneficis* took its place in the devils' compendium.[85] Along with the first printing of the Frankfurt 'Faust', the *Historia von D. Johann Fausten*, this marked the first high point of an international wave of demonological literature, which culminated in the works of famous writers such as the Swiss Reformer Lambert Daneau, the French legal theorist Jean Bodin or King James VI of Scotland / James I of England (1566–1625), in whose *Demonology* the weather witches played a major role.[86] Apart from their practical side, demonologies sought to provide a secure foundation for human knowledge, beyond devilish illusions; they thus addressed a central epistemological problem of their age.[87]

The pact with the devil was one of the favourite themes around the year 1600. The first appearance of the Faust legend in print in 1587 was followed by several other versions. Soon it was being translated into English (1588), Dutch (1592) and French (1598). Christopher Marlowe (1564–93) wrote a drama on the subject and had it performed in London. In the same year as the first Faust book, the Jesuit Jacob Gretser (1562–1625) brought out his version of the

143

pact with the devil, *Udo of Magdeburg*, and four years later Jacob Bidermann (1578–1639) created another Faustian figure in his *Cenodoxus*.[88] As Goethe clearly recognized, the figure of Faust pointed beyond the usual witch craze: moreover, he sold his soul not for base material gain but to achieve greater knowledge of nature. He therefore has a place in the origins of modern natural science.

The Zeitgeist is also reflected at the level of belles-lettres, which are usually classified under epochal categories such as Renaissance, Mannerism or Baroque.[89] 'Fleeting time, transience, a gaze focused on the end: such ideas and phenomena are often to be found. This devaluation of the temporal stands face to face with a passionate striving for the eternal. People wallow, as it were, in worry, sorrow and fear. . . . In a spirit of resignation, weariness and indifference to life and what it brings, gained through personal experience, renunciation, humility and a longing for inner peace highlight the idea of participating as actors and spectators in a theatrical performance, which is life itself. This all points to a devaluation, a conscious belittling of the external goods of life.'[90] The embodiment of this Zeitgeist may be found in the prose of Miguel de Cervantes (1547–1616) or the plays of William Shakespeare (1564–1616), while Andreas Gryphius (1616–64), who knew Holland, Italy and France from his travels, was quite overcome by the distress of his age. In strongly metrical sonnets such as 'Vanitas Vanitatum Vanitas', this syndic of the Silesian Estates described the wanton destruction of his times, whose continual crises made all human planning appear pointless. Cities are born today, ruined tomorrow; the human visage, created in God's image, 'is tomorrow ash and bone'.[91]

The intense descriptions of winter, a pendant to the wintry landscapes in painting, deserve special attention. Simon Dach (1606–59) writes, for example: 'Now the mountains and fields are asleep, covered with frost and snow, and the forests too have hidden in their coat of white. The rivers stand peaceful and still which would otherwise be sweetly flowing on and away.'[92] Of course it has been argued that in Baroque literature, as in pictorial art, we are talking only of standard tropes.[93] And actually such a dealing in types does correspond to the poetics of the Baroque, as suggested by Martin Opitz (1579–1639) and others.[94] If we were to group such passages alongside one another, many phrases would appear interchangeable. In his 'Auff die nunmehr angekommene kalte Winterzeit', for example, Johann Rist (1607–67), a preacher from Wedel in Schleswig-Holstein, penned the following: 'Winter has arrived / Snow covers the land / Summer has flown away / Forest has turned into frost. /

4.7 The triumph of death and the vanity of all earthly things were typical themes in an age of insecurity. Michael Wohlgemut's *Imago mortis* – from *Liber Chronicarum*, Nuremberg 1493.

Meadows are disabled by rime / The fields gleam as metal / The trees are changed into ice / The rivers stand like hard steel.'[95]

Can one rationally interpret such texts without knowing that, in the cold winters of the Little Ice Age, rivers such as the Rhine or Rhone did repeatedly freeze all the way down to their bed? The same applies to the threat of sickness and death. Yet Wolfram Mauser, to take but one example, sees no biographical reference in a precisely dated poem that was actually presented as autobiographical: '*Thränen in schwerer Krankheit. Anno 1640.* I know not what ails me / I sigh all the time, day and night I weep. / I suffer a thousand pains and fear a thousand more. / The strength fails in my heart / My spirit pines away / My hands droop down. / My cheeks grow pale / The sparkle fades from my eyes / As the glow from a burned-down candle. / My soul is bombarded as a lake in March. / What else is this life . . . / than a dream mixed with bitter fear?'[96] For historians it seems obvious that writers in that age of crisis were referring to their contemporary situation when they wrote thus of their fears.[97]

The Cool Sun of Reason

The need for a new order

The experience of insecurity and disorder was as symptomatic for early modern Europe as it is today for underdeveloped countries.[98] Not for nothing do modernization theorists see European societies of that time as the test case for their theories;[99] its disasters and the cultural reactions to them were the consequence of a relative lack of development and security. Early modern Europe also provides us with an impressive example of a way out of climatically determined emergencies. As it curbed scapegoating and found a rational way of dealing with climate change, Europe drew on its own strength to find an egress from the vale of tears. The keyword for this liberation from premodern 'sin economics', which pushed religious delusions into the background and adjusted to the living conditions of the Little Ice Age, was *reason*.

Discussions of famine and undernourishment in the twentieth century have shown that agrarian societies were prone to crisis not only because they had to face adverse climatic conditions (drought, excessive rainfall, cold or heat), but also because specific cultural factors created a series of further difficulties: unsuitable landholding structures (such as large estates or 'feudal' rents), unsuitable political structures (inadequate public institutions, administrative corruption, minority use of the state as a tool of appropriation), unsuitable economic structures (shortage of capital, the degree of market integration, poor storage and agricultural organization), a deficient system of education (low literacy levels, poorly trained elites, inadequate instruction in agriculture and nutrition), a deficient health system (scanty hygiene, inadequate medical care) and problems with communication infrastructure (poor knowledge of resources and prices, inefficient transportation of food over land and water) and with risk insurance.[100]

The struggle for greater stability as a crisis-handling strategy was a fundamental tendency in early modern Europe.[101] This need for order had far-reaching consequences. The formation of strong core states under monarchical or parliamentary aegis made sense during the development of an economic system that Immanuel Wallerstein

146

has called the 'European world system', which incorporated the newly created global network of colonies.[102] State-building was a reaction to the sense of insecurity in a period of increased tensions, while a standing army was meant to hold internal and external threats in check. A monarchical order was better than religious civil war – that is one way of reading the *Leviathan* by Thomas Hobbes (1588–1679).[103] This legitimation of the strong state reflected a common European tendency to establish clear boundaries within and without, to guarantee external defence and internal order. But the quest for order was not only a process implemented 'from above'; it also corresponded, for all the fractiousness, to a need 'from below'.[104] One of the tasks of the bureaucracies founded in the seventeenth century was to improve the infrastructure by building roads and canals, to develop an effective educational system, and to promote better storage methods, medical care and hygiene.[105] Another element in the stabilization of modern state systems was the development of the political sciences, natural, international and public law as subjects to be taught in the universities.[106]

As Norbert Elias understood, these institutional changes did not come without a price attached to them. The inculcation of social discipline, through the army and the state administration, was one part of this. But, even without physical compulsion, the new institutions placed new demands on people. In the innermost circles of the central ruling apparatuses – in the royal courts, for example, but also in parliament or the collegiate-like public authorities – expertise, elegance and psychological refinement played a greater role than muscular strength. Self-discipline or the internalization of new values here operated on a spontaneous basis, and the motivations for it were more complex than expected by Norbert Elias.[107] Among these may be included a consciousness of sin in a religious society reeling under the impact of various disasters. The developing structures of Europe's international trade and communications were also good reason for procedural rationalization and self-discipline. Many details of life became subject to regulation in a society based on an ever tighter division of labour. This was as true of state functionaries as of people living in care homes or members of guilds and fraternities. Even travellers were affected: instructions to state employees in the eighteenth century underlined the values of punctuality and accuracy on professional grounds alone.[108]

The need for security and order led to normative endeavours in more and more areas of life. Like people living at the time, so may we too take astronomy as the emblematic science – one that Johannes

Kepler (1571–1630) cast into mathematical laws, in accordance with the neo-Platonic criteria of 'world harmony'.[109] Geometry helped in the development of norms of conduct, by supplying a universal model of regularity that could be applied to anything from dancing to transport networks, from state administration to fortress architecture and horticulture.[110] The regulatory systems that were sought and partly founded in the seventeenth century were the answers of new generations to old passions. Lastly, the reforms even affected language. Literature oriented itself to classical poetics, with its firm rules in verse and drama, as well as to the great rule-generating institution of the age, the Court.[111]

The rollback of religious fanaticism

The disenchantment of the world fed on a number of sources, which acquired a deeper meaning in the context of the Little Ice Age. In the early modern period, society began to emancipate itself from the supremacy of religious thought. Perhaps it is only at first sight surprising that this process began in the age of religious wars and witch persecutions. For the turning away from religious fanaticism, the secularization of human thought, may be understood as a direct response to that experience. The catalogues of the Frankfurt Book Fair show a rise of non-religious factual literature in parallel to the flood of works reflecting theological disputes. In the catalogues of Georg Willer, for example, it is precisely in the hunger year of 1570 that a new category first appears alongside theology, law and medicine: *Philosophici, artium liberalium atque mechanicarum, & libri miscellani*, which showed that the sciences and humanities and the academic technical book had achieved a new dignity.[112] In the section dealing with works in German, rather amorphously headed 'Various Books in all the Arts and Other Fields', we find books on the weather and agricultural reform. And the first year's catalogue already mentions a new farming manual by Konrad von Heresbach (1496–1576),[113] which was subsequently reprinted four times and translated into English.[114] In the following years, the book fair offered for sale a number of Italian and French works on the theory and practice of agriculture.[115] And in the 1590s came the standard work of 'paterfamilias literature' by Johannes Coler (1566–1639), *Oeconomia ruralis et domestica*, which went through numerous editions right down to the eighteenth century.[116] All these works offer recommendations for the improvement of crop rotation and the application of fertilizer, and point in a direction that would lead in the Netherlands and England to the revolution in agriculture.[117]

It is not always clear from the title of meteorological works whether they belong to the genre of spiritual edification literature. Often, in fact, they are a hybrid form combining natural observation with futurology,[118] or a chronicle of the times with theological interpretation.[119] Many preachers, like the Tübingen professor Johann Georg Sigwart, duly specialized in the correct theological interpretation of extreme climatic events.[120] The astronomer Bartholomeus Scultetus (1540–1614) ventured a kind of long-range weather forecast in his *Prognosticon von aller Witterung*. The narrowly meteorological works of the seventeenth century, as well as its commentaries on Aristotle, could do with closer examination. That this might prove rewarding is shown by Leonhard Reynmann's often reprinted *Wetterbüchlein*, which avoids any metaphysical explanation in terms of God, the devil or witches.[121]

The age of suffering in the middle of the early modern period led to durable learning processes. The associated process of rationalization is usually put down to long-term structural causes: to the rise of capitalism in Werner Sombart, of Protestantism in Max Weber or of the Absolutist court in Norbert Elias. But perhaps we might also say that the religious fervour wore off. It seemed evident to some people at the time that holy wars and witch hunts improved neither the weather nor the harvest but merely caused additional suffering.

The publishers who shied away from theological disputes and preferred to concentrate on factual literature included such illustrious figures as Matthäus Merian in Frankfurt and Johann Carolus (inventor of the periodical newspaper) in Strassburg. The example of Carolus shows the extent to which the seventeenth-century media constituted a new agenda: miracle tales and the fervour of competing religious denominations had no place here. The weekly editions of *Relation* in 1609 reported in detail on Rudolf II's edict of tolerance for Bohemia, the *Böhmischer Majestätsbrief*, and the ongoing crisis in Jülich-Kleve. The medium was not only the message; it also conveyed one, namely, that the world was awry. And, to understand this world, people needed news from the realm of politics rather than metaphysics.[122]

In this context, the fencing in of passions with the help of the pragmatic philosophy of neo-Stoicism[123] made as much sense as the natural science of Galileo Galilei, whose method was communicated to quite a broad public,[124] or Francis Bacon's controlled experimentation, which was meant to standardize knowledge and to ensure the methodical progress of science.[125] The aversion of dogmatists to this kind of open-ended research was already apparent before the Inquisition moved against Galileo. For this reason, the statutes of the Florentine Accademia dei Lyncei explicitly barred theologians from membership; scholars wished to discuss serious problems freely, without censorship or sterile religious quarrels. The Inquisition's action against the leading scientist of his age lit the beacon. All later private or public scientific societies – such as the Royal Society in England or the Academie Française in France – excluded theologians from their deliberations. A legend grew up in the Protestant countries that no science was possible where the Pope's authority held sway. In the seventeenth-century visions of utopia – Francis Bacon's *New Atlantis* as much as Johann Valentin Andreae's *Christianopolis* – theologians were always assigned a minor role, while philosophers were entrusted with the steering of the ideal state. In reality, after the Peace of Westphalia (1648) and especially after England's Glorious Revolution (1688), religious pressure and episcopal despotism rapidly subsided.[126]

Scientific revolution and the incipient optimism of progress

In searching for early forms of modern science, one finds sixteenth-century researchers driven by a Faustian thirst for knowledge for whom the boundaries of experiment and magic were unclear.[127] In the eyes of theologians, *Magia naturalis* was a highly dubious discipline. Like the artist Benvenuto Cellini (1500–71), many magi dabbled in the exorcism of demons.[128] But natural magicians were held in high regard at the courts of the age. Leading examples such as Girolamo Cardano (1501–76),[129] whose serious researches are still echoed in the driveshaft and other inventions, or the Neapolitan Gianbattista della Porta (1535–1615) represented natural science better than the universities, whose curricula did not extend to the investigation of nature (apart from medicine, astronomy and mathematics).[130] The English magus John Dee (1527–1608), noted for his experiments and predictions, was similarly in demand at the courts of Elizabeth I in England and Rudolf II in Prague.[131] The occult sciences, including astrology, Paracelsian medicine and alchemy, were a passion of the age, which was not content with old answers to new problems and struck out in directions of its own.[132]

Not only Galileo but already Cardano showed that scientists lived dangerously because of potential conflicts with religious orthodoxy.

150

As with Cellini, the religious authorities had good reason to mistrust Cardano, della Porta and Dee, since their experiments sometimes left open the possibility of being interpreted as magic rituals. Cardano found the appropriate answer, however, in his preoccupation with mathematics, to which 'Cardano's formulas', as they are known, still testify today. Kepler understood how to present his central astronomical statements in mathematical formulas, which still have validity as 'Kepler's laws' of planetary motion. With Galileo the natural sciences opted for a language against whose formal logic no church could take exception: the language of mathematics. The introduction of natural laws disempowered the personal God of the Bible, who could arbitrarily haunt the countryside together with angels and demons. *Laws of nature* – even when attributed to God's perfection – now bound the creator to rigid rules that could be scientifically investigated.

After Galileo Galilei and Francis Bacon, this decisive paradigm shift excluded the religious and spiritual world from the store of explanations. So ended the discursive supremacy of theologians and magi, alchemists and astrologers. The great renewal of the sciences in Bacon's *Instauratio Magna* fed on a regularization of 'wild thought' through the standardization of experimental procedure and the systematic discussion of results. This permanently shifted the limits of knowledge and offered the vision of systematic scientific progress. It was no longer church or state but a worldwide community of philosophers and scientists that decided on the truth.

In the early seventeenth century, the natural sciences began fundamentally to change our picture of the world. The mathematician René Descartes (1596–1650), with his philosophical division between mind (*res cogitans*) and matter (*res extensa*), epistemologically denied to spiritual beings (angels, demons, God) any influence on the material world. The mathematician and physicist Isaac Newton (1643–1727), with his law of gravity, found a kind of formula valid in the far reaches of the universe as well as here on earth, and thereby demonstrated that the laws of nature were generally applicable.[133]

In this way, Newton fulfilled an old yearning of the magi to join together microcosm and macrocosm conceptually. In 1703 Newton, who had a presence in every branch of science – astronomy, mechanics, optics and acoustics, arithmetic – was elected life president of the Royal Society for the Advancement of Learning, which had devoted itself for half a century to systematic study of the laws of nature. Because of his scientific achievements, he was also elevated to the peerage and chosen to represent the University of Cambridge in

Parliament. For enlightened intellectuals, scientists such as Newton were the true heroes of history; they had carried forward the whole of humanity. It was also especially alluring that scientific and political progress were so closely bound up with each other in the English case. Freedom from censorship promoted freedom of the sciences. So, it seemed more than pure chance that Newton's key work, the *Principia Mathematica*, had appeared more or less simultaneously with the Glorious Revolution of 1688, the deposition of the English monarch and the beginning of parliamentary rule.[134]

Although the natural science of antiquity was heavily attacked from the sixteenth century on, it was only now that its supremacy in the universities collapsed. Instead of Aristotelian theory, with its elements, humours, substances, qualities and causes, Galileo and Hobbes inaugurated a *mechanical philosophy* that Newton subsequently raised on to the altar of human thought. Its new picture of the world was well suited to cope with specific theoretical and political problems.[135] Observation of nature began its long period of ascent, with the help of understandable and repeatable experiments. At the same time, scientists investigated especially puzzling phenomena such as magnetism and electricity, for which (as for gravity) there had previously been only magical explanations. After the founding of specialist journals such as *Philosophical Transactions*, the results of experiments no longer had to be communicated by letter but could be printed and made available for anyone to read.

The new sciences developed not only new institutions (scientific societies, academies) but also their own media, in which international discussion of experiments could proceed without religious or secular interference. This proved extremely effective in rapidly exposing errors or fakes, since spectacular experiments, such as the ascent of hot air balloons, could be imitated within weeks around Europe and North America and either confirmed or dismissed. Experimentation became one of the favourite leisure pursuits of the upper classes of society. The creation of new 'locations' or institutions for scientific work, respected media and institutionalized channels of communication led to that 'paradigm shift', in the sense of a new consensual model, which Thomas S. Kuhn (1922–96) famously attempted to describe.[136]

The invention of the telescope in the early seventeenth century and of the microscope at the century's end firmly established the use of instruments in the natural sciences. In the eighteenth century, optical devices and measuring instruments became part of the regular equipment of all academies and laboratories, and also of many an exalted private home. Barometers, thermometers, hygrometers, theodolites, pumps and prisms were produced, bought and used in large quantities. Even quite banal measurements found their way into print – for example, the daily atmospheric pressure readings in the *Gentleman's Magazine*.[137] Although most experiments served the purpose of entertainment, or at best of elementary research, people did also actively and successfully aim to come up with new inventions. In 1751 a preoccupation with electricity spurred the American scientist Benjamin Franklin (1706–90) to invent the lightning conductor, and in 1782 an abiding interest in the chemistry and physics of gases led to almost simultaneous invention of the hot air and the hydrogen balloon, and thus to the possibility of air travel. They were seen as the crowning achievement of the Age of Enlightenment – as signal victories of science and technology over religion and superstition. Surprisingly, this balance sheet rarely features inventions that were of even greater significance for economic development and the transformation of man's natural environment: those which, in the wake of the Industrial Revolution, eventually had effects that are today blamed for climate change.

The rule of the 'Sun King'

After the major wars of the seventeenth century, political stability came to be guaranteed by standing armies. The greater distance of rulers from ruled, which found architectural and ceremonial expression in magnificent buildings, increased the sense of security and created new space for the sciences within nation-states, such as the city-states of the Renaissance had never been able to provide. The establishment of standing armies made society not more just but more secure and predictable, by stabilizing its existing relations and reducing the threat of revolt or revolution from the lower orders. The age of political unrest was over; the ideological charge of interdenominational rivalry was defused.[138]

But the rigours of nature, the age of cold and infertility, were not yet under control. The 'Sun King' construct, with its power of metaphor, therefore served to shore up the promise of a better future. Not only Louis XIV of France (1638–1715, r. 1643/1651–1715) but also political opponents such as Holy Roman Emperor Leopold I (1640–1705, r. 1658–1705) stylized themselves as heat-providing central stars.[139] The reign of the 'Roi Soleil' and his officials witnessed not only the rise of France to great power status but also the especially severe cold waves of the *Maunder Minimum*.[140] In those grim years,

4.8 The young Louis XIV of France, personifying the sun in ballet costume. Rival monarchs also had themselves represented as sun kings in the Little Ice Age.

the coldest of the last millennium, serious harvest failures brought shortages and starvation, as well as epidemics with high mortality rates. It is not as if there were no other problems. For decades France, like Austria, waged one war after another, frittering away the country's wealth and piling up debts that would burden the state budget for years to come. But the caprices of the weather added to the insecurity. In the winter of 1683–84, frosty weather lasted from mid-October until Easter of the following year.[141] Especially inclement years, such as those of the mid-1690s,[142] give us a picture of how large parts of Europe – from the Swedish province of Finland[143] to the south of France – were plagued with terrible food emergencies.[144] Even in the last years of the Roi Soleil, officialdom was incapable of effectively alleviating the hardship. Relief efforts pointed in the right direction, but they became entangled in demarcation disputes between the provincial and central authorities. The only difference with the past was that starvation and despair did not reach the main political centres. The Lyons city council provided a model of how to use all available means to escape a hunger crisis.[145]

And there was one other difference. In the century of the Enlightenment, famines were increasingly understood to be the result of mismanagement. The public was no longer prepared to accept sermons about divine retribution, but pointed to the structural deficits and political omissions that hindered relief operations after crop failures. Why were the roads so bad that bread cereals could not be expeditiously imported? Why were the warehouses too small to supply the poor? Why were the existing ones not full? Why were the harvests too sparse, and why had royal officials not made adequate provision? The criticism of intellectuals and the anger of the populace made rebellions more likely and forced the authorities to act. Although drastic measures were usually taken only against profiteering grain merchants and bakers, it was perfectly obvious to the government that it could ensure its survival only by making better provisions. In principle, things were no different in the civilizations of antiquity: dire straits for the population called the legitimacy of rulers into question. The examples of England and the United Provinces of the Netherlands show that things could work out well where there was effective parliamentary control. The Enlightenment increased the pressure on governments to carry out reforms.

Some reform projects of the sixteenth and seventeenth centuries had an effect in the medium term. In the eighteenth century, the building of roads and canals was stepped up, so as to expand the range of transport options for items of mass consumption; and the introduction of coach networks placed national and international long-distance communications on a completely new foundation. With what then counted as a modern postal service, it was possible to learn at once of incipient regional emergencies and to comb Europe or further afield for resources that might fill the gap. Extensive and well-organized commercial links helped to soften the impact of supply crises. Maritime trade made it easier to organize relief, as well as to improve storage, administration and health provision. It was thanks to these efforts among others that old scourges of humanity such as the plague disappeared from Europe. The *agrarian revolution* that began in the Netherlands in the late sixteenth century,[146] and which England took up with a slight delay, tore up the evil at the roots. Dyke-based land reclamation, moorland farming, crop rotation, irrigation and the sowing of new seed varieties helped to ensure that famines became a rarer occurrence. All this happened after 1709 in the space of a single generation. The improvements in infrastructure and food production created a solid basis for more rapid urbanization, which in turn reduced people's sense of dependence on nature and its powers.[147]

As the distance from primary production grew, people became less susceptible to superstition and religious convulsions. The wealth of the upper classes created an effective antidote to the rigours of nature. We can see this, of course, in the great courts of Versailles, London and Vienna, or among aristocrats and wealthy merchants, who were no longer afraid of witchcraft; but it is especially apparent in the newly prosperous Netherlands, which as the chief trading centre for cereals became the motor of world trade during the Little Ice Age. Here, not only had witch-hunting long been abandoned; even heretics and Jews had the possibility of a relatively good life.[148] Even the harsh winters seemed less threatening. In painting, the icy landscape turned from an apocalyptic vision of horror into a display of joy or composure in winter sports. Hendrik Averkamp and many others actually specialized in winter scenes, which were produced in thousands of variations for the homes of rich citizens.[149]

The extreme winter of 1739–1740 as a test for the Enlightenment

The spirit of the Enlightenment, which prevailed in the courts and academies, was increasingly marked by an optimistic faith in the rationality of nature, the advance of the sciences and the perfectibility of society, and even by the idea that human beings were naturally good and were corrupted only by a bad upbringing. Philosophers such as Jean-Jacques Rousseau (1712–78) assumed that all people were equal and should have equal rights; this included religious and social minorities and also – a great novelty – women. All Enlightenment thinkers were agreed that the aim of education should be to make people capable of thinking for themselves.

The church, not the state or capital, was generally regarded as the power of darkness; it kept people in ignorance, stoked up fanaticism and bore responsibility for countless atrocities – from the Crusades through anti-Jewish pogroms and witch-hunts to the extermination of indigenous Americans and the wars of religion. To religious fanaticism the Enlightenment counterposed the ideal of religious toleration, one shared by many kings and emperors such as Joseph II (1741–90, r. 1765–90). The thinkers of the age even had many supporters in the ranks of educated theologians and prince-bishops.[150]

However, although the fate of the ruling classes was no longer really at the mercy of climate fluctuations, the same was not true of ordinary people. Precisely when disasters struck, therefore, conservative preachers were tempted to rouse the populace against the Enlightenment political elite. Since Europe was largely agrarian and depended for food on the success of its harvests, the weather still usually played a decisive role. Only in Britain and the United Provinces, with their milder ocean climate, had agricultural improvements and infrastructural development gone far enough to remove any major dependence on regional crop yields.[151] Elsewhere in Europe, each harvest failure due to bad weather became a test for the Enlightenment. In a sense, then, the decades after 1709 were a stroke of luck, since the moderate climate fostered illusions that life was getting better. Probably it was a little colder than it is today. But the cool sun of the Age of Reason created stable living conditions, which accorded with the optimistic philosophy of Christian Wolff (1679–1754) and its doctrine that the world rested upon a divine 'pre-established harmony'.[152]

It is customary to regard the Lisbon earthquake of 1755, when a tsunami devastated the Portuguese capital, as the test case for Enlightenment optimism. But, in the context of this book, we should rather take the extreme winter of 1739–40, which once again led to a major subsistence crisis on the continent and to high mortality rates in many areas. This cold wave was debated in the academies and in learned journals, the periodical press and numerous published works.[153] These emphasized again and again the 'natural' causes of the extreme climate, which resulted in firewood shortages, hypothermia, under-nourishment, disease and increased mortality.[154]

In 1740 Johann Rudolph Marcus, a pastor from Mühlstädt, drew a balance-sheet of previous periods of disastrously cold weather and offered a kind of history of the Little Ice Age from a Saxon point of view.[155] He reported numerous examples such as burst thermometers, smashed bridges in Amsterdam, frozen wine in cellars or ink in inkpots. Even in Russia it was colder than usual, and wild animals froze to death in the forest. In Persia people died of cold. In Scandinavia all the lakes and rivers iced up. No windmills worked any longer, and industrial production ground to a halt. Anyone who had to go outdoors faced a difficult time. The postillion and his horse froze on their way from southern Holland to Amsterdam; the Hamburg courier arrived dying on his steed. The postillion and his passengers froze en route to Berlin, as did people in their own homes in Poland. In Scotland, too, many starved to death,[156] and in France and England there were deadly influenza epidemics.[157] In Paris and the provinces, 'many people died from flu and colds. It is calculated that from the beginning of this year until May more than forty thousand were buried in Paris, and that in the month of April alone more than four thousand died in its hospitals. It is thought that such epidemics, illnesses and sudden deaths were the effects in their bodies of the excessively harsh winter, although in other times the cold allowed the pestilence to take hold.'[158]

In recent decades, the Great Winter of 1740 and the crisis of the early 1740s have attracted considerable interest as a research topic.[159] From both comparative overviews and local microstudies,[160] we now know that the effects were lesser than in similar previous crises. At the first sign of harvest failure, many governments began to stockpile cereals from other regions, including some far beyond Europe. England made purchases of wheat and rye as well as rice in the Russian Baltic, Ottoman Egypt and its own North American colonies. Prussia had hoarded so much in its granaries that Friedrich II – after his attack on Austrian Silesia in breach of international law – could distribute seed to the peasantry to win their loyalty. Prices shot up in 1740, and households had a rough time of it because of their extra spending on bread. But hunger became synonymous with bad, irrational government. There was a dramatic rise in mortality only in marginal areas such as the Scottish Highlands or Ireland.[161] Seen in this light, the extreme winter of 1740 was not only a test case but also a triumph for the Enlightenment.

High smoke and great fear

The inclement weather of the 1780s was due to the eruption of Laki and other volcanoes in Iceland and Japan. In the Tokugawa empire Mount Asama, 150 kilometres from the capital Edo (today's Tokyo), exploded in the spring of 1783, causing some 35,000 direct victims in the densely populated surrounding area. The long-term consequences were even more disastrous: the skies grew completely dark in Edo itself; and the emissions led to widespread soil contamination, heavy rainfall and major countrywide cooling that lasted for several years. This volcano-induced climatic change resulted in serious failures of the rice harvest. The price for this staple food tripled, and there were terrible famines in the years between 1783 and 1787. Poorer sections of the population tried to feed off roots and nuts, cats and dogs; there were even cases of cannibalism. The regular censuses show that the starvation years must have claimed several hundred thousand lives, as the total population fell from c. 26 million to c. 23 million within a few years, whereas it had otherwise been fairly constant throughout the century.[162] Against this background, peasant uprisings and armed resistance broke out in the 1780s, reaching a climax in May 1787 around Osaka and Edo as well as in northern Japan. Peasants and city-dwellers joined together to destroy the houses of privileged merchants and profiteers. It seemed 'as if the world was out of joint'. After the death of Shogun Ieharu (1737–86, r. 1760–86), his government was overthrown because people blamed the disasters on his corrupt liberalization of the feudal system. The situation then rapidly improved under the new shogun, Ienari (1773–1837, r. 1786–1837). The country returned to feudal structures, as the terrible years through which it had lived were interpreted as divine retribution.[163]

With these Japanese events in mind, it is interesting to consider how Europe was reacting at that time. The eruption of the Laki volcanic fissure in Iceland began in May 1783 and lasted eight months over a length of 27 kilometres. It hurled vast quantities of gases, volcanic ash and aerosol into the stratosphere and led to persistent darkening of the sky. The 'high smoke' was observed in Copenhagen on 29 May, in Paris on 6 June and in Milan on 18 June.[164] The emissions caused sulphurous odours, eye irritation, breathing difficulties and headaches as far away as Central Europe. In large parts of Europe and the Ottoman empire, there were reports of thick 'dry fog' and darkening or unusual colouring of the sun. Members of the Societas Meteorologica Palatina reported that, in the summer of

1783, the dark sky meant that people were able to look at the sun with the naked eye. In Iceland acid rain turned large areas of land into waste, destroying crops and making cultivation impossible for years to come. Acid rain damaged the environment throughout Scandinavia. Vegetation directly suffered even in the Netherlands, where with some delay cold and drought led to harvest failures, outbreaks of fever and diarrhoea, and increased mortality. In Iceland itself some nine thousand people – a quarter of the population – died as a result of the Laki eruption.[165]

Many people at the time saw the darkening as an evil omen, a harbinger of apocalyptic changes. Enlightenment periodicals, guided by science, tried to combat these superstitious viewpoints. Swiss researchers measured the weakening of the sun's rays, and in Saxony it was discovered that a magnifying glass could no longer produce the melting of lead seal. A connection was already drawn between the 'high smoke' and the heavy rainfall or crop damage of that summer.[166] As a result of the communications revolution, the 'high smoke' and general cooling were described and discussed in scientific journals; comparisons were drawn and explanations suggested. The Jena professor of mathematics, Johann Ernst Basilius Wiedeburg, made the assumption that similar phenomena had preceded the cold years of 1709 and 1740.[167] The American scientist, inventor and communications entrepreneur Benjamin Franklin, who was then American ambassador to the French court, was the first to establish the link between volcanic eruptions in Iceland and stinging in the eyes – his own eyes, in fact, at a considerable distance away.[168]

In the 1780s we find a rash of extreme events such as we saw before in the Maunder Minimum. These have recently been associated with the Laki eruption. Grain prices rose threefold in the decade beginning in 1784. The cumulative periods of cold during these years led to heavy snowfall and deep frost, widespread failures of vine and bread cereal harvests, flooding and livestock epidemics – precisely the combination of disasters that hits traditional agrarian societies hardest. The severe winter of 1783–4 saw exceptionally heavy snowfall, and there was serious flooding when the thaw came in late February. In the Rhine-Main area, the highwater mark at that time has in many places never been exceeded; it caused devastation in the fields or meadows and outbreaks of livestock disease due to contamination of the land. Many bridges collapsed, and roads and paths became impassable. The winter of 1784–5 was also exceptionally long and cold. In Berne snow lay on the ground for 154 days, whereas in the cold winter of 1962–3, for example, the corresponding figure was only 86 days.[169]

The French Revolution

The effects of the climate phenomena depended on the social and cultural context. Staple food prices rose worldwide after the volcanic eruptions of 1784, and France was no exception. But, unlike in Japan, there was no relaxation of tension in Europe and North America, since the wet and cold weather was followed by abnormally dry conditions. The drought of 1788 impacted differently on Tsarist Russia and on an Enlightenment France then in the final phase of feudalism. In the United States, European settlers who had shaken off the colonial yoke looked courageously to the future, whereas liberation movements developed among the indigenous peoples, who saw the hard times as the fruit of repression by the new American nation.[170]

Many factors played a role in the outbreak of the French Revolution: long-term and short-term, political and cultural, economic and social. The long-term structural crisis is the level at which Jack Goldstone thought he could locate the cause of revolutions. In just the twenty years before the Revolution, the population of France grew by two million, or ten per cent.[171] Since 1770 agricultural output had no longer kept pace with the demographic trend: the agrarian sector was still very traditional in comparison with England, and English travellers such as Arthur Young thought of it as backward. Farm buildings were of poor quality, grain storage capacity was inadequate, and the situation was bad in relation to irrigation, fertilizer and fodder; the traditional three-field system was practised as in the time of Charlemagne; and all agricultural labour – sowing, harvesting, threshing – was done by hand. The peasantry was generally hostile to innovation.[172] All this resulted in food shortages and an increased susceptibility to crisis. The structural crisis first made itself felt as part of the great European hunger of 1770, in which malnutrition and epidemics were also rife in Germany, Switzerland and other parts of continental Europe. The end of the 'golden age' of Louis XV (1710–74, r. 1715/23–1774) was already a period of great hardship.

Under Louis XVI (1754–93, r. 1774–92), whose rule opened with another hunger crisis, the situation came to a head. The whole economy began to stagnate, and a recession developed in 1778 as a result of the fall in purchasing-power. The liberal economic policy

161

was a decisive factor in the escalation of the crisis. By 1785, owing to its aggressive policy of free trade, the monarchy no longer had food stocks at its disposal to cushion the effects of harvest failure. And in 1787, on top of everything else, the regime gave free rein to grain exports that benefited the nobility. This boosted profits for grain producers but intensified the shortages for consumers. A loss of purchasing-power by as much as 50 per cent further increased the demand for cheap industrial products from England, which resulted in lay-offs and unemployment in the French handicrafts sector. In the wake of these policy failures, industrial, agricultural and social crises broke out together in 1787–8.

This explosive cocktail was compounded by a cluster of phenomena characteristic of the Little Ice Age. The year 1788 witnessed a severe drought, and in the middle of July a hailstorm devastated large areas of land. The British ambassador Lord Dorset reckoned that 1,500 villages were seriously damaged in the vicinity of Paris. The grain harvest yield throughout France was twenty per cent down on the average for the previous decade. Prices kept rising for more than a year until the outbreak of the Revolution. The hail-ravaged summer was followed by the harsh winter of 1788–9, when economic life came to a standstill. The spring thaw led to flooding and consequent livestock epidemics. Hunger revolts broke out in many regions. Whole families attacked and confiscated grain convoys, sometimes paying a price they considered 'fair' that did not correspond to the current market situation, sometimes simply plundering the food. Military escorts were deployed to protect the consignments, and in the towns there was widespread fear of mob pillaging and aggressive bands of looters.

The growing number of beggars led to major tensions. The problem was less the traditional poor, who qualified for alms handouts, than beggars who came from outside the area. Rumours of so-called 'brigands' caused such fear that weapons were given to the peasants to defend their land. This phenomenon has gone down in history as *la grande peur*. The renewed drought in 1789 added to fears that the situation would worsen. Rivers ran dry, and watermills – the driving force of industry – ground to a halt; the result was flour shortages and higher prices for bread.[173] The climate cannot, of course, take all the blame for the outbreak of revolution in France, but the historian Ernst Labrousse (1895–1988) has shown that hunger was one of its main causes and that the urban and rural masses who helped it to triumph had suffered the most from food shortages.[174] One statistic epitomizes this nexus between revolution and hunger. Grain prices reached their highest level on 14 July 1789, the day of the storming of the Bastille.[175]

The Tambora freeze, democratization and cholera

The most interesting case of volcanic cooling followed the eruption of Tambora on 10–11 April 1815, which has been classified as the most violent eruption of the past ten thousand years.[176] It caused global cooling of 3 to 4 degrees for a number of years, the so-called *Tambora freeze*.[177] The powerful explosion sent huge quantities of ash and aerosol into the stratosphere, and over the coming months these were distributed all around the world. The year 1816 was known in Europe and North America as the 'year without a summer'. Among the Tambora effects were unusual phenomena in the sky, the disappearance of sunshine in Iceland, hunger crises in many European countries, waves of emigration, harvest failures in the USA and India (where there were also outbreaks of disease),[178] a drought in South Africa and dramatic witch persecutions in the kingdom of Zulu ruler Shaka (1789–1828, r. 1816–28) and elsewhere in southern Africa.[179] In Europe and the United States, the main response to the crisis was the kind of technocratic-administrative management that has become so typical of Western civilization. Governments and officialdom in the German territories kept their distance from the hungry masses and suppressed disorders or acts of assault, if necessary with military force – as in the case of the anti-Semitic Hep-Hep riots in 1819.[180]

Pre-industrial pauperization should be looked at again in the light of climate history, which might show that the formation of non-peasant, sub-bourgeois layers was not least the result of a traditional, climate-induced cycle of harvest failure. With the experience of the French Revolution and the revolutionary wars at the back of their mind, European politicians reacted sensibly to the outbreaks of unrest. The parliamentary system in England, the Declaration of Human Rights, the independence of the democratically constituted United States of America, the French Revolution and finally the French Constitution of 1814: this whole history led the bourgeoisie in Central Europe to reject a return to absolute monarchy. Hunger-related disturbances following the Tambora freeze helped to delegitimize the 'cabinet governments' in small states such as Baden, Bavaria and Württemberg and gave an impetus to opposition forces.[181] There was thus a link between the climate and the reform movements of 1818–20 that led to constitutions with a strong role for parliament, to what is known in German as *Frühkonstitutionalismus*.[182]

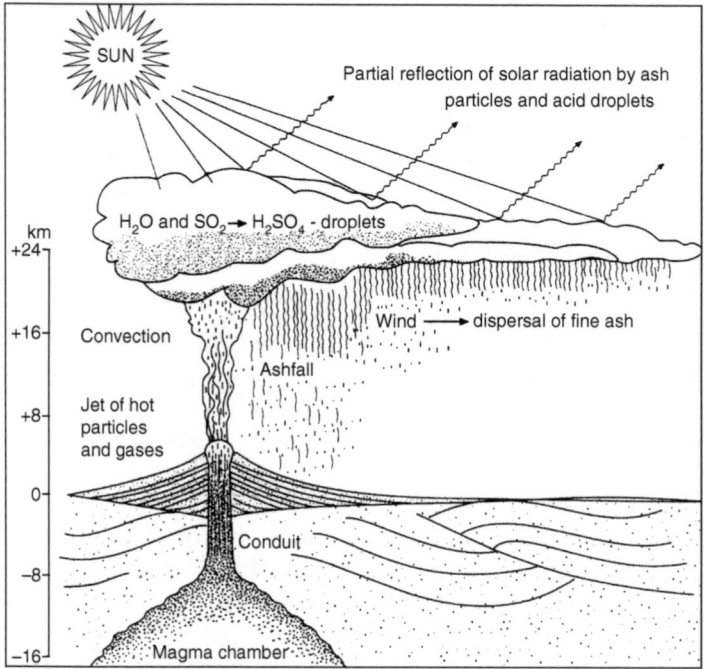

4.9 The Tambora eruption in 1815 thrust gases and aerosol into the strato-sphere, causing worldwide cooling. In the next two years there were major social and political crises on every continent.

In the same way that the hunger periods of the 1330s and 1340s preceded the Great Plague in Eurasia and North Africa, so did the Tambora freeze promote another distinctive epidemic: cholera. This is caused by the *Vibrio Cholerae* bacterium, which leads to severe diarrhoea with loss of body fluid. The initial outbreak was in the Indian subcontinent, where the disease had long been endemic in the Ganges delta.[183] Then, in the years after the Tambora freeze, it spread through the Russian empire to Europe, where it had hitherto been completely unknown. Contributory factors in Europe and North America were the high level of urbanization and poor hygiene condi-tions among the lower classes that were comparable to those in large Indian cities. Since the cholera agent settles in the intestinal flora, sanitary conditions, impure drinking water and deficient wastewater drainage were important elements. Pauperism and malnutrition also played a role.

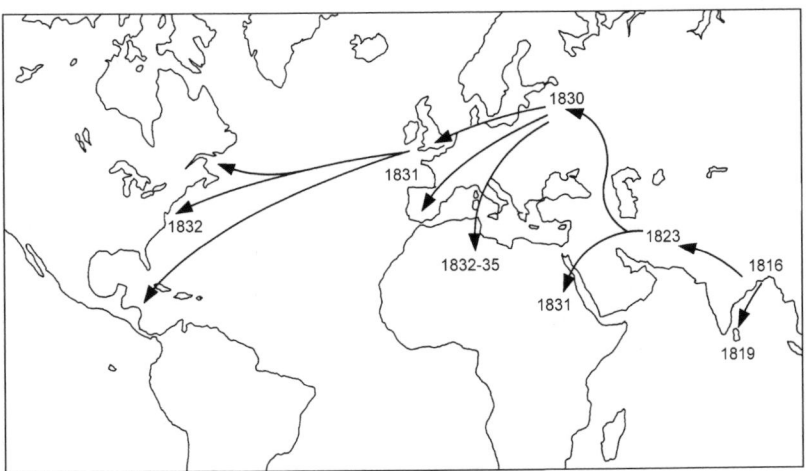

4.10 The first global cholera pandemic, in the wake of the Tambora cooling. It spread from India through Russia to Europe, and from there to North America.

The cholera epidemic had a major impact. In Warsaw alone 2,600 people died of it, in Berlin roughly 1,500 (0.6 per cent of the population), including the philosopher Georg Wilhelm Friedrich Hegel (1770–1831). In Paris it claimed 18,500 lives, or roughly two per cent of the population. When the rumour spread that poisoned food was to blame, there were attacks on producers of food and drink. From Hamburg the outbreak probably crossed the North Sea to London, in 1832.[184] Unlike the cities of continental Europe, the British capital already had its own drinking water supply network, so that mortality remained quite low in many districts. In 1832 the disease reached North America, where it raged until 1848 and claimed more than four thousand lives in New York alone.[185]

Hunger, emigration and revolution

A varied diet reduces vulnerability to harvest failure, whereas monoculture can have devastating results. The best-known example of this is the Irish famine of the 1840s.[186] Other countries too were hit by the arrival of the *Phythophtora infestans* plant pathogen, probably through guano imports from South America. In continental Europe the fungus that triggered the potato blight did not survive a period

165

of drought, and moreover potatoes were there only one staple food among others; it was possible to compensate for the harvest shortfall by importing and boosting the consumption of wheat and rye products, rice and maize. For Germany, a whole series of studies has demonstrated the success of official policies to handle the crisis, although it is also clear that the crisis mood played a role in the period leading up to the 1848 revolution.[187]

The famine assumed dramatic dimensions in the Scottish highlands and, as in Ireland, unleashed a wave of emigration to the United States.[188] The special character of the Irish famine was due to the fact that the potato harvest was 30–60 per cent down in several successive years (1845, 1846, 1847, 1848), in a country where it was the main staple. People managed to cope with the first hunger year, thanks to aid from the British government under Robert Peel (1788–1850, r. 1841–6), which had purchased large quantities of cornmeal in the United States as a precautionary measure. But the second harvest failure exhausted the supplies; people started slaughtering livestock and selling their possessions. There was also a change of government in London. Committed to a free trade policy, Prime Minister John Russell (1792–1878, r. 1846–52) held the view that regulation of the crisis should be left up to the market. The food supply subsequently collapsed in Ireland, and poor sections of the population died en masse from starvation, fever, gastric and intestinal disorders, typhus or scurvy. Mortality was so high that in late 1847 the British colonial regime introduced free soup kitchens, even though supporters of Charles Darwin tried to explain away the deaths by natural selection. According to recent estimates, Ireland suffered a population loss of approximately 25 per cent – more than in all earlier epidemics in the history of the island.[189] No other event has been as deeply engraved in the Irish – and, because of emigration, the American – collective memory as the Irish Great Famine.[190]

For many European countries, the Revolution of 1848 brought reforms that were limited to civil liberties and better parliamentary representation. In Japan, on the other hand, a slightly later cool period led to a complete system change. As before in 1783–7 and 1833–9, there were dramatic harvest failures and hunger revolts in the years between 1863 and 1869. Cold waves and heavy rainfall, which curtailed the rice harvest, were probably triggered by a super-El Niño.[191] At the same time, conflict sharpened with the Western powers, which sought to end Japanese isolationism and to open up its ports. External pressure, necessitating mobilization and higher taxation in Japan, combined with widespread urban and rural uprisings to bring about the collapse of the Tokugawa regime. In many parts of the country, people withheld taxes and rejected conscription notices. In major cities such as Kobe, Osaka and Edo (Tokyo), rioters stormed government buildings, destroyed the homes of moneylenders, and burned land registers and promissory notes. The last military dictator, Yoshinubu (1837–1913, r. 1866–8), had to abdicate in 1868 at the age of just twenty-one. After months of civil war, the government fell into the hands of the underage Meiji Tenno (1852–1912, r. 1868–1912), in whose name the empire was reformed in line with Western models.[192]

— 5 —

GLOBAL WARMING: THE MODERN WARM PERIOD

Apparent Uncoupling from the Forces of Nature

The agricultural revolution

Contemporary discussions have tended to locate the causes of global warming in industrialization. At first, however, industrialization had quite a different significance: it promised to free people from hunger by ostensibly uncoupling them from the forces of nature. Until then agricultural yields had been so low, and general prosperity so limited, that it had scarcely been possible to pay the increased food prices that followed harvest failure. Hygiene conditions and living accommodation were also meagre, public health services still in their infancy. The result was a great susceptibility to disease, with high mortality in all age groups. Average life expectancy, taking into account infant and child mortality, was still only thirty years in Europe around 1700 – less than it is today in the poorest developing countries.[1]

Several active encroachments in nature were crucially important in improving the food situation. First, the landscape was again visibly changed in order to gain additional arable land and to ensure constant irrigation. The draining of great marshland areas such as the Oderbruch began in the eighteenth century. The flow of major rivers was controlled – the Rhine, for example, with its endless bends and dead arms, had kept large areas of land out of service and made it impossible to regulate agriculture. In the late nineteenth century these efforts culminated in great dam projects, which symbolized man's victory over nature. Artificial lakes were meant not only to save drinking water and to prevent floods, but also – after it was discovered how to use electricity – to generate electric energy on a grand scale. These innovations were, of course, characteristic of all industrial countries. As the historian David Blackbourn has stressed, the changes to the landscape were tantamount to a second conquest of nature.[2]

Second, new farming methods and food or fodder plants, together with crop rotation and artificial fertilizer, drove up the efficiency of agriculture. Science gained entry to agriculture through the application of biology and agrarian chemistry. Soil was deliberately improved, while animals and plants were bred to adapt to their local area and produce higher yields. Crops from Asia or the Americas, such as maize, rice and potatoes, had already been introduced in the early modern period. But their cultivation was stepped up only between 1750 and 1900. The significance of bread receded as diets became more varied, with the result that prayers such as 'Give us this day our daily bread' are scarcely intelligible in today's Europe. Third, new technologies and machines were for the first time applied to agriculture on a large scale. This made it possible to farm ever larger areas in an ever shorter time. Even the fast-growing population of the second half of the nineteenth century could be fed without any problems.[3]

Hygiene and healthcare

Better nutrition was the surest route to higher life expectancy and therefore to population growth. And, since economic theory in the seventeenth and eighteenth centuries equated population growth with economic strength, it became one of the goals pursued by the state. The systematic use of quarantines since the seventeenth century had already helped to contain all kinds of epidemics and to reduce the incidence of disease. Another important area was hygiene. According to the miasma theory of disease, it was toxic vapours that caused people to fall sick. Such emanations appeared wherever putrefaction or decomposition processes occurred – for example, in marshes or any kind of filthy place. Avoidance of dirt was thus a question not only of civilization[4] but of survival. Since the late Middle Ages, many towns had kept their streets clean and regularly emptied pit privies, but general paving and regular street-sweeping were enforced only in the eighteenth century. Improved house construction also helped to keep epidemics at bay. After the Great Fire of 1666, the plague never returned to London, since the worst areas were burned down and the new metropolis was built in stone.

Another important point was the improvement of medical care. Here, though, we are talking only of a tendency, since in the state of medicine at the time it could be positively dangerous to be treated by a doctor. As the causes of most diseases were unknown, specific therapies remained at best ineffectual. But doctors did recommend more suitable lifestyles and advised against excessive drinking and eating, as well as overuse of alcohol and tobacco; they also urged sick patients to stay in bed and gave advice on general care and – perhaps most important – hygiene. The slowly rising number of doctors per head of the population may be taken to indicate a growing awareness of these problems.[5] In addition, empirical studies of the causes of disease became more common in the Age of Enlightenment, and the results were discussed in newly founded specialist journals. New institutions also came into being to treat and investigate illnesses: hospitals.

The hospital is a typical invention of the Enlightenment. There were, of course, leprosy and plague houses in the high and late Middle Ages, which served only to isolate sufferers rather than to treat them. The 'hospitals' of the late Middle Ages were not what we understand by the term: they admitted only chronically sick people whose infirmity was not contagious. Perhaps the first large hospital in the modern sense was the one founded in Vienna, in 1784, by Emperor Joseph II; it examined and treated sick people from all over the city and its surroundings. The Vienna *Allgemeine Kranken-Haus* (AKH) existed until the 1990s, when the city-centre complex was closed because of its antiquated building standards and replaced with a hospital of the same name a few kilometres away. The criticism that the AKH had served more to study diseases than to cure the sick refers to its dual function: treatment took place there within an academic context. Its aims thus include research and medical training, with a potential to expand medical knowledge and to improve methods of treatment. In addition to inpatient and outpatient care, hospitals also have the functions of aftercare and obstetrics.

Improvements in nutrition, hygiene and healthcare, together with changed patterns of marriage, led to population growth unparalleled in human history. The industrial countries entered a period of demographic transition[6] as birth-rates increased, mortality fell, and life expectancy moved upward. The outcome was the kind of demographic explosion that we can see today in countries of the Third World. It furnished the youthful industry with so much manpower that, in line with the law of supply and demand, wages remained low and goods could be produced relatively cheaply. Labour market conditions permitted rapid industrial expansion and the accumulation of large fortunes in the hands of a fast-growing bourgeoisie. The bourgeoisie of the industrial age was different from the old European burgher class, which had been largely dependent on state authorities of every kind (monarchy, city governments, guilds and corporations). The industrial bourgeoisie was more independent, wealthier and more self-assured. It had an influence over the respective national culture and politics – for example, in the constitutional parliaments or during the 1848 Revolution.

The Industrial Revolution and fossil energy use

The term 'Industrial Revolution', which first appeared in the 1840s, denotes a fundamental transformation in human history. Many compare its importance to that of the 'Neolithic Revolution', the transition from hunting to agriculture on the threshold of the Late Stone Age. In both cases we speak of a revolution because the developments in question, though initially local, spread around the world and changed human life so radically that a return to earlier conditions was no longer conceivable.

Even before, however, industry did not consist simply of individual crafts. People had been familiar with industrial windmills and watermills since the high Middle Ages; in the textile sector there were stamping-mills, fulling-mills, woad mills, tanning-mills and silk-thread mills; or, in metal production, edge mills, grinding-mills, wire-drawing mills and forge hammers.[7] In the English language, indeed, the word 'mill' came to stand for a factory.[8] In the industrial mills, but also in mining and manufacturing, hundreds of people were already employed before 1750. Nevertheless, the Industrial Revolution marked a qualitative leap in the development of production, since productivity shot up and a greater division of labour made it possible to produce goods more cheaply. Industrialization brought the beginnings of the modern 'consumer society', in which even ordinary people can eat and clothe themselves properly.

Again it was Britain, the motherland of freedom of thought, that saw a successful transfer from the sciences to the economy. We have already noted England's pioneering role in the 'agricultural revolution';[9] now the same was happening in industry. The Industrial Revolution marked the transition from agrarian to industrial society, since an ever higher share of the population was employed in industry. English entrepreneurs such as Thomas Newcomen (1663–1729) had been experimenting with steam engines as early as the beginning of the eighteenth century, and since being patented by James Watt (1736–1819) in 1769 artificially driven machines had gained entry into industry. The new steam engines could run on wood. But, since the forests in the British Isles had been largely cleared in antiquity and again in the early modern period for ship-building, it was initially coal that served as the main fuel for the steam engines; it also had the advantage of being able to generate higher temperatures. The wood shortage meant that coal was also employed for domestic heating. Thus, systematic use of fossil ener-gies was already a feature of the seventeenth century. By the time that industrialization began, people in England had long been accus-tomed to the smell of burning coal, although it was only now that the process of mining accelerated. Coal or coke was also used in iron-smelting, and increasing quantities of iron and steel were needed for the new industries.[10]

Contrary to a widespread misconception, the Industrial Revolution did not start with steam engines but only later allied itself with this new breakthrough. The leading sector in England was textiles, and especially the cotton industry, which had first been introduced in 1585 when some weavers from the Spanish-besieged city of Antwerp emigrated to England. The raw materials were imported from coun-tries in the Mediterranean. But production remained so small in the seventeenth century that most of the cotton fabrics sold in London originated in India. Once cotton began to be grown in America, however, a law was passed in 1700 banning textile imports from Asia. At first cotton fabrics were produced in the putting-out system, whereby entrepreneurs distributed the raw material to weavers, who then conducted the work on their own account. This gave rise to a developed division of labour typical of the proto-industrialization period.

The production process began to change when John Kay launched his fly shuttle in 1733. Also in the 1730s a spinning-jenny was invented which, in combination with other machines, could be driven by engine power. Its designers thought that the machine would be able to run on horse, water or wind power, and that it would work to best advantage in a 'mill' or factory. The factory buildings now built by Richard Arkwright (1732–92) became the birthplaces of modern mass production. By 1789 there were already 2.4 million spinning-machines in England, nearly all driven by water power for reasons of cost. The great number of machines deployed in textile production led to a huge demand for standard-ized wood and metal parts, which gave rise to a machine industry in its own right.[11]

The steam engine as universal motor

James Watt acquired his first patent in the same year as Richard Arkwright, although nearly twenty more years passed before the first steam engines found their way into the Arkwright mills. In the intervening period, Watt's invention had developed into the general-purpose steam engine. But in the eighteenth and early-nineteenth centuries most factories continued to operate on natural water power: this was certainly cheaper, but another reason was that until the 1820s watermills made from wood remained more productive than steam engines. After 1800, though, steam engines were more widely deployed, partly because the growth of the metallurgical industry made them cheaper and more reliable.

Since the days of Thomas Newcomen, the main use for steam engines had been in coal-mining, where until the late seventeenth century water was still being pumped out by muscle power in the shape of horse-driven gins. Newcomen's steam pumps were much cheaper to run and soon established themselves in the mines. By the year 1780 nearly a thousand of Newcomen's steam engines were in use. Their higher consumption of coal was understandably not a matter of concern to colliery owners, since the material on which they ran – unlike horse feed – was anyway at hand. Nor were they interested in the better use of energy that Watt's rival steam engine offered. For lack of demand, Watt had to file for bankruptcy, but he was rescued by the metalware producer Matthew Boulton (1728–1809). The firm Boulton & Watt discovered the owners of tin and copper mines to be customers in need of coal. The breakthrough came only when Watt succeeded in making the motion of piston rods circular. This proved the way to an all-purpose engine.

Mechanization spread around 1800 to the production of iron and finally to mining. Nuts, bolts and cogs, tools, bars and rails were needed in ever larger quantities and could no longer be produced by manual labour. A special machine tool industry took into account the specialization in metal processing. Mass production made it necessary to increase transport capacity, first through the building of canals, but then through a complete revolution in transport. The age of the mobile steam engine dawned – which meant steamboats on water and railways on land. Production of these means of transport, together with increased weapons production, multiplied demand for iron and fuel and provided the second great impetus to industrialization. The new transport facilities made possible a huge increase in textile exports, and therefore in textile production.[12]

The Industrial Revolution placed England at the centre of the world economy. Industries soon sprang up elsewhere: in Scotland, Belgium, northern France, the Rhineland and Westphalia, Switzerland, Saxony and Silesia, Bohemia, the Vienna Basin, Hungary, northern Italy and the northeast of the United States. These old European and American industrial regions are still today among the centres of the world economy, even if the circle of competitors widened in the twentieth century.[13] The rise of coal-fired steam engines began in the first half of the nineteenth century. By 1838 the number of steam engines had grown to more than three thousand. At the same time their power increased from 10hp to 50hp, and textile factories gradually converted to the new form of energy. Owners of textile factories now covered 75 per cent of their power requirement from steam engines and were their leading customers, followed by mines, foundries and the iron industry. Demand also appeared from the Continent, so that Boulton & Watt found itself supplying customers throughout Europe and in India. Yet all European countries combined had fewer steam engines than England, which in these decades rose to become the 'workshop of the world'.[14]

Around 1880, industrial emissions probably still had no effect on the global climate, or at least none of any significance in comparison with 'natural' variations. The Little Ice Age was heading for another 'cold maximum', all the more intense because of the volcanic eruption on Krakatoa in Indonesia, which was soon known of throughout the world thanks to rapid advances in telecommunications. The cooling led to smaller harvests in 1884. But in the industrial countries this was scarcely noticed because of the general production surpluses.

Coal use in the railway age

From the outset, industrialization had a great influence on the environment, though at first this was confined to nearby regions. Apart from deforestation, the main effects came from the production of solid, liquid or gaseous waste and the generation of noise. Industry also required land for factories, roads, railways, canals and workers' dwellings. These developments were independent of the political system: whether France was an empire, a monarchy or a republic was of no significance for its consumption of raw materials and the environment.

Whereas in Britain the Industrial Revolution began in the textile sector, the central place on the Continent was occupied by railway construction; this gave other European countries their first opportu-

World coal production[16]

year	mn. tonnes of coal
1860	132
1880	314
1900	701
1920	1193
1940	1363
1960	1809

5.1 Worldwide coal production.

nity to close the gap with England. The railways required less land than roads or canals, they were not dependent on the weather, and it was possible to operate them according to standard schedules. In 1825 the first steam train began to run between Stockton and Darlington, and in 1830 the great industrial cities of Manchester and Birmingham were linked up. It was not long before the railway spread to the Continent: the first German service, between Nuremberg and Fürth in the Kingdom of Bavaria, opened in 1835, and the Saxon cities of Leipzig and Dresden were linked up in 1839. From the middle of the century, the various stretches of track grew into a veritable network, and this led to nationalization of railroad companies that had previously relied on private funding. Soon the network was expanding across borders, from Belgium to Italy. The highpoint of railway construction was reached between 1870 and 1910, under state management.

The creation and running of railway systems initially depended on coal-fired steam engines. Annual coal consumption in England was approximately 11 million tonnes in 1800 and twice that figure in 1830, but the railway age considerably accelerated the expansion. By 1870 consumption was running at 100 million tonnes per annum.[15] Production on the Continent also rapidly increased with the mining of deposits in Belgium, France and Germany, and later in Poland and Russia.

The rise of oil

Meanwhile oil had joined coal as the second fossil fuel. Already in antiquity, where the Greek word was *petroleum*, it had appeared on

175

the surface in many places and was used for a number of purposes. In ancient Mesopotamia it served to caulk ships; in China it was used for lighting and in medieval Europe as an ingredient in medicines. In 1857 Bucharest introduced it for street lighting. And two years later the American industrialist Edwin Laurentine Drake (1819–80) conducted the first economically significant drilling in Titusville, Pennsylvania, which led to an oil fever and use of the fuel in industry. In the 1890s, when the spread of electric lighting threatened to make it marginal once again, another use for oil was discovered: its refinement into petrol.

The invention of the combustion engine by Nikolaus Otto (1832–91) gave a permanent basis to the demand for oil. Combustion engines gradually replaced old-fashioned steam engines, the main impetus for this coming from the invention of the motor-car by Gottlieb Daimler (1834–1900) and Carl Benz (1844–1929) in the late 1880s. Mass automobile production began in 1908 with Henry Ford's Tin Lizzy. After the introduction of conveyor belts, production of this model reached no fewer than 15 million units by the year 1927. In the United States, the leading oil producer, more cars were driven in the first half of the twentieth century than in the whole of the rest of the world. Many car producers emerged in Europe too, but all of them together never reached Ford's production figures. Tin Lizzy's record was finally matched only in 1972, half a century later, by the Volkswagen Beetle.

The 1950s as the threshold of a new epoch

As we can see from the car production figures, fuel consumption only gradually increased in the course of the twentieth century. A qualitative leap came in the 1950s, when oil replaced coal as the primary fossil fuel. This is clear from many indicators, such as the load capacity of oil tankers: since the 1930s not only the number of ships but each one's capacity had increased many times over, from roughly 20,000 tonnes to some 400,000 tonnes. The soaring demand for oil went together with the discovery of more and more uses for it. The petrochemicals industry produced artificial materials that had a place both in homes and in manufacturing processes. It was in the 1950s that plastics began to take over from traditional packing materials such as paper, wood, glass or metal. One of the reasons for the oil boom was the production of heating oil, which after the Second World War increasingly took the place of coal. Oil was cheaper to produce and cleaner to process, as well as having numerous applications. But its main use remained the production of fuels.

The military-strategic importance of oil became evident in the First World War, when motor vehicles, tanks and airplanes were deployed for the first time. This fanned the interest of the great powers in overthrowing the Ottoman empire, on whose territory most of the Middle Eastern oil deposits were located. In the 1950s, after the discovery of abundant reserves in the Gulf region, Washington installed regimes in Iran, Iraq, the United Arab Emirates and Saudi Arabia that would comply with the interests of US oil corporations by granting them licences. It did not shrink from overthrowing freely elected presidents such as Mohammed Mossadegh (1882–1967, r. 1951–3), because he had questioned the oil-prospecting monopoly of the Anglo-Iranian Oil Company (later renamed British Petroleum, BP) and engineered the nationalization of such activity. Oil wars have since been a recurrent feature of the region: one has only to think of the three Gulf wars and the overthrow of Iraqi dictator Saddam Hussein, on the transparent pretext that he possessed weapons of mass destruction.

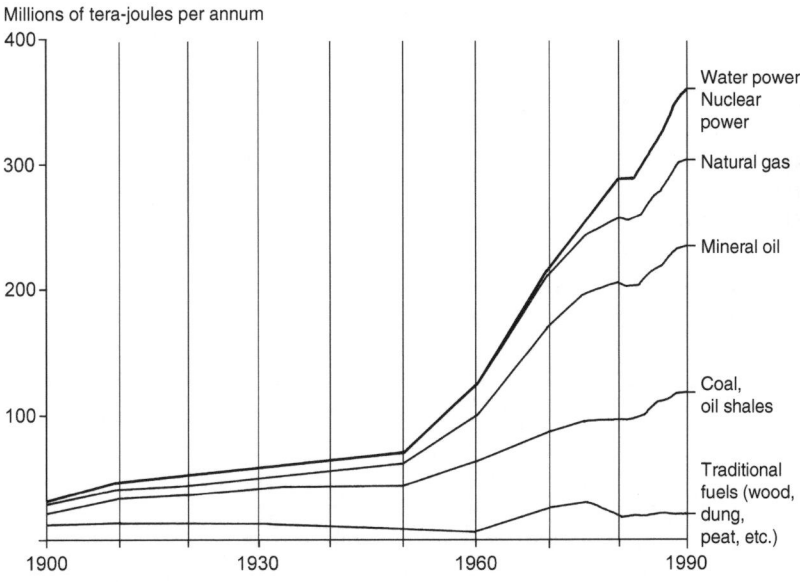

5.2 The evolution of world energy consumption 1900–1990.

The 1950s also saw the beginning of the boom in civilian air travel. It is true that companies like Lufthansa had already been founded in the 1920s, but flying had remained the preserve of a small upper class. Only when it became normal practice in the business world did it come down in price and become accessible to much wider layers. The arrival of mass tourism in the 1970s brought a rapid increase in the numbers of passengers and destinations as well as the capacity of individual aircraft. Energy consumption per passenger rose disproportionately, however, as an aircraft holding three hundred passengers needs as much fuel as tens of thousands of Volkswagen Beetles.

Annual energy consumption increased sixty times over between 1860 and 1985, but the greatest leap in demand occurred in the second half of the twentieth century. According to the Mannheim environmental scientist Jörn Sieglerschmidt, the 1950s represented an 'epochal threshold' in the relationship between human beings and the environment.[17] For several decades it looked as if it might be possible to escape the old constraints of economics and ecology. Cheap energy led to urban sprawl, the spread of industry into new areas, a nationwide supply of consumer goods and the opening of the countryside to mass motoring. Christian Pfister speaks of an 'environmentally historic crossing of the threshold from industrial to consumer society'. Energy-devouring goods began to flood into private homes: electric cookers and washing-machines, refrigerators and toasters, chest freezers and microwaves, dishwashers and vacuum cleaners, electric irons and electric toothbrushes, hood dryers and portable hairdryers, radio and television sets, record players, cassette players and video recorders, computers, printers and scanners – every home became a machine park. And the electrification drive continued into cellars, garages and hobby rooms: drilling equipment, screwdrivers, compass saws, lawnmowers, hedgecutters, etc. In each room a radiator and an array of lamps: all this was unknown at the beginning of the twentieth century and, even if the technology already existed, was still unknown for most people before the 1950s.[18] Until the first oil price crisis of 1973–4, energy-saving played no role in machinery or domestic appliances. Only after the shock of that OPEC price hike did serious discussion begin about low-consumption engines, alternative energies, heat conservation, and so on.

The oil boom caused environmental problems to mushroom as emissions spread around the globe. The negative effects of urban sprawl and pollution soon became visible. But at first no one thought of a much graver consequence: the burning of coal, oil and gas was releasing chemical elements that had been deposited deep in the earth millions of years ago, during the 'carbon epoch' in the history of the planet. The energy sources were not minerals but former life, organic material – or, in the evocative words of environmental historian Rolf Peter Sieferle, 'underground forest'.[19] In the burning of wood, carbon is released from the current carbon cycle and then bound again in the growth of plants. In the burning of fossil fuels, however, carbon is released that was bound 300 million years ago and deposited in the earth when the carbon forests died. This additional carbon, once released, can never be bound again in the growth of vegetation, but combines in the atmosphere to form compounds such as the trace gas carbon dioxide, CO_2. Since the 1950s we have been witnessing an exponential enrichment of the atmosphere in carbon dioxide, methane (CH_4) and nitric oxides. The production of CFCs (chlorofluorocarbons) only began in the 1950s. These trace gases contribute to the warming of the earth's atmosphere and are therefore known as 'greenhouse gases'.[20]

The population explosion

In order to evaluate how humans have historically influenced the climate and environment, it is not enough to focus on the evolution of change-producing trace gases in ice core samples. One significant indicator is population trends, since the use of land and energy depends on them. Estimates of world population in history have become increasingly precise. For the eve of the Neolithic revolution c. 10,000 years ago – that is, the end of the time when virtually all humans lived as hunter-gatherers – Carlo Cipolla sets a lower limit of 2 million people and an upper limit of 20 million who could have lived in habitable areas of the world. For the period around 1750 he works on the assumption of a total of 750 million. This population of three-quarters of a billion represents the historical peak of the agricultural phase of human development. For the year 1850 the world population has been estimated at 1.2 billion; and industrial food production would open up more and more new growth possibilities. In 1900 there were 1.6 billion people living on earth. For 1950 the UN's *Demographic Yearbook* sets the figure at approximately 2.5 billion, and by 1975 it had already climbed to 4 billion.[21]

In the year 2000 the earth's population was a little over 6 billion. But the highest ever growth-rate was in 1970: more than 2 per cent globally. At that time a world total of 12 to 15 billion was being predicted for the year 2050, but it was also then that a number of developing countries began to introduce a conscious demographic

179

policy. In the country with the highest population, the People's Republic of China, the Communist Party ordered a one child per family policy to be introduced in the 1970s, and Chinese birth-rates have since fallen by two-thirds (from 3.0 per cent to 1.2 per cent). Demographic growth has also been restricted in other countries of Asia and Latin America. In Africa alone does it still continue as before, held down only by diseases such as malaria and AIDS. In the industrial countries of Europe and North America, despite growing immigration, population figures have been falling slightly. The cutback of global birth-rates to 1.4 per cent, with a tendency for them to decline further, has led the United Nations to revise its population forecast for the year 2050 to a total of 9 billion.[22]

Critiques of civilization and discovery of the limits to growth

The critique of industrialization is as old as the phenomenon itself. In some cases – such as the machine wreckers of the eighteenth century or the Communist and socialist movements – it has been closely associated with a critique of undesirable trends in society; in other cases, such as Romantics or anarchists, it has led on to a more far-reaching critique of civilization.[23] One thinks here of the 'barefoot prophets' of the early ecological movement, who before the First World War and again in the 1920s dropped out of the social mechanisms of industrial society in protest against the exploitation of nature and human beings.[24] When the modernization paradigm held sway, man's impact on the environment increased to such an extent that industrialization, deforestation and environmental pollution became issues of global importance. With the space travel of the 1960s, the first photos of Earth from space dramatically opened people's eyes to the fragility of our biosphere within the vast expanse of the universe. The moon flight was thus of great significance for the development of ecological awareness.

The most important challenge to an unbridled optimism of progress came in the early 1970s from the 'Club of Rome' group of scientists, politicians and economic leaders, which had been founded in 1968 on the initiative of the Roman industrialist Aurelio Pecceri (1908–84).[25] A team of scientists under Dennis L. Meadows presented a report to the Club in 1972, using numerous data and projections to define the 'limits to growth'. They identified five trends in the global system and analysed their interaction with one another: accelerated industrialization, rapid population growth, worldwide

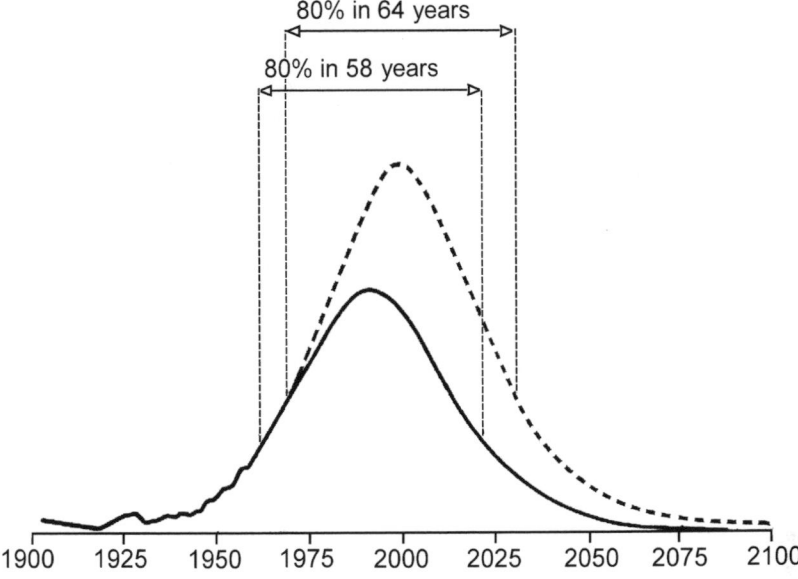

5.3 The oil age as it appears in M. King Hubbert, *The Energy Resources of the Earth*, 1971.

undernourishment, excessive use of reserves of raw materials, and increasing destruction of habitat. The authors emphasized that global resources were finite: the reserves of coal and oil deposited 300 million years ago in a long geological process would be used up within two hundred years. Once burned, they would be lost for ever.

The policy implications were substantial. In the eyes of the Club of Rome, it was necessary to limit growth and to create a different kind of society characterized by social justice, human rights and a new harmony between man and nature. The reporters were unable to say, however, how this could be achieved under the sign of economic liberalism and a total of some two hundred national states each pursuing their own interests. The conclusion was clear: 'Every day of continued exponential growth brings the world system closer to the ultimate limits to that growth. A decision to do nothing is a decision to increase the risk of collapse.'[26] In common with all critiques of civilization, the Club of Rome report was marked by a degree of social utopianism. The novel element was that it came from the mainstream of society, from a group of successful scientists, businessmen and politicians.

181

Today, the limits to growth debate leave us with mixed impressions. On the one hand, the pace of industrial growth has increased even further since 1972, as more and more countries in Asia and Latin America have turned into modern industrial societies; world population growth has slowed, the food situation has improved, and environmental protection has become established policy in the developed countries. But the overexploitation of nature continues, and raw material reserves are being used up at an even faster rate. Since new deposits keep being discovered, the end of modern civilization has not appeared on the horizon. Paradoxically, the thawing of permafrost and the melting of sea ice actually opens up fresh possibilities for the prospecting of fossil fuels.[27] In retrospect, the Club of Rome seems to belong in the ranks of false prophets who predicted disasters that never came to pass. Yet it remains the case that reserves of raw materials – especially of fossil fuels – will at some point be exhausted. The merit of the Club of Rome is to have driven home the point that natural resources are finite. Greenhouse gases were not an issue in those days.

The Discovery of Global Warming

Early greenhouse theories

Warnings of a new ice age continued in the 1960s, but then a decade later global warming was discovered – or, to be more precise, rediscovered. For, early in the nineteenth century, the French physicist Jean-Baptiste Joseph, Baron de Fourier (1768–1830) had raised the question of what accounted for the temperature of the planet. He appreciated the significance of the earth's atmosphere, which he compared to a kind of greenhouse that somehow retains part of the heat from the sun's rays. The so-called greenhouse gases were discovered in 1859 by the Irish physicist John Tyndall (1820–93), who established that the sun's rays penetrate oxygen and nitrogen, the main components of the atmosphere, but that the longer heat waves returning from the earth are held back by carbon dioxide. It was on this basis, he concluded, that the earth could heat itself and support life. Tyndall already entertained the notion that this might explain climatic variations in the history of the planet; he was thinking especially of the great ice ages, which had just been discovered and were the subject of heated debate. In Tyndall's view, a fall in the amount of water vapour in the earth's atmosphere had caused a lessening of the greenhouse effect and thereby precipitated the ice ages.[28]

In 1896 Svante August Arrhenius (1859–1927), the future Nobel prizewinner for chemistry, drew attention to the problem of rising CO_2 emissions in the wake of industrialization. The Stockholm professor, too, was more interested in explaining the ice ages than in murky predictions about the shape of things to come.[29] Shortly before, the English geologist James Croll had developed a theory of 'feedback', according to which glacier expansion led to further cooling by increasing the radiation of sunlight back into the cosmos; this in turn led to a change in winds and ocean currents. Croll had discovered the albedo effect and come to grips with a complex climate model. As soon as an ice age began, it kept itself going in various ways. Arrhenius worked out what would happen if the CO_2 content of the atmosphere were to be halved, his conclusion being that temperatures would fall by 5 degrees Celsius. This may seem little, but a downward spiral could be set in motion through Croll's feedback effects.[30]

In an age before the invention of computers or even calculating machines, such calculations involved a lot of tiring paperwork. Arrhenius asked a colleague whether major changes in atmospheric

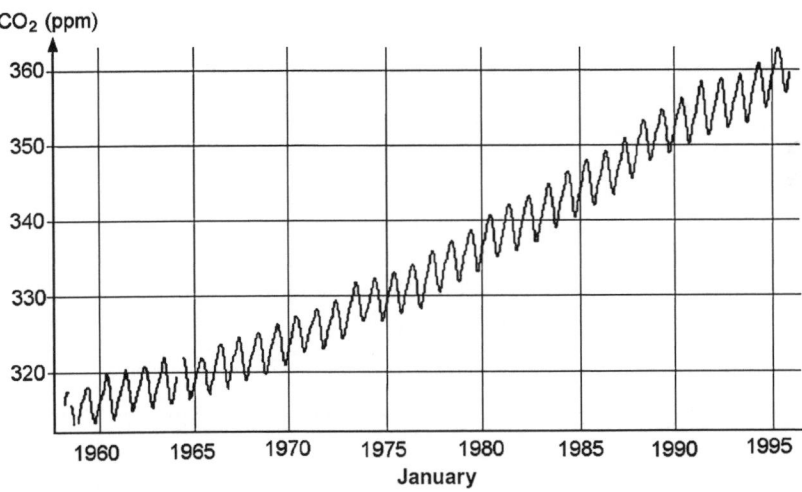

5.4 The Keeling Curve is a climatological icon, which for the first time clearly represented the carbon dioxide enrichment of the atmosphere. Since the 1960s it has been regularly updated with new measurements.

composition were even possible. Arvid Högbom, who had begun to calculate the CO_2 emissions from coal-burning in factories and homes, thought they did not add up to much in comparison with naturally occurring carbon dioxide, but he believed that they would inevitably build up over time. Arrhenius also did some calculations and concluded that a twofold increase in atmospheric carbon dioxide would lead to a rise of 5 to 6 degrees Celsius in the earth's temperature. This prospect did not trouble him at all; a little more warmth would not come amiss in Scandinavia. Besides, he expected such a change to occur over a period of at least several thousand years. He shared the technological optimism of his time, which thought that resourceful engineers would surely find the answers to all problems and create a better, fairer future. What Arrhenius had outlined was a theoretical curiosity, a thought experiment. He did not consider himself to be the discoverer of global warming.[31]

At roughly the time when Arrhenius was doing his calculations, the Little Ice Age came to an end; the Thames froze one last time, then a general warming set in. Meanwhile Milanković was developing his explanation of climate changes in terms of astronomical orbits. In the 1930s, when drought was turning western states of the American High Plains into a dust bowl, people also became interested in the trend; Guy Steward Callendar again took up the question of global warming.[32] In 1956, with the help of a climate model, Gilbert Plass for the first time formulated a carbon dioxide theory of climate change.[33] On the occasion of International Geophysical Year in 1957 – which gave a lasting boost to geological and climatological science – Charles Keeling published measurements covering many years that would earn him considerable fame. His purpose in doing so was to demonstrate the annual variations in atmospheric CO_2 and thus, in a sense, to visualize the inhalation and exhalation of the biosphere in a yearly rhythm. The measurement base, Mauna Loa in Hawaii, was far away from big cities or continental masses and therefore promised to deliver a picture undisturbed by environmental pollution. Rather unexpectedly, however, the yearly rhythm was affected by a second phenomenon: a constant upward trend in the concentration of CO_2 in the atmosphere.[34]

At the beginning of Keeling's series, the concentration of CO_2 was approximately 315 ppm (parts per million), then it grew to 325 ppm in 1970 and to 335 ppm in 1980. The measurements have since been regularly repeated, in order to be sure that the atmosphere has continued to become enriched with this greenhouse gas. The readings have constantly risen: from 360 ppm in 1995 to a record 380 ppm

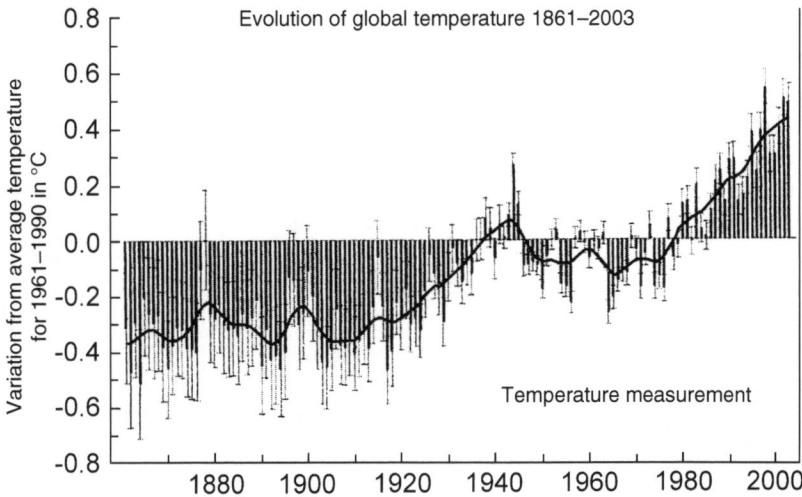

5.5   From the Little Ice Age to global warming. Since the 1890s temperatures have shown an upward trend, although it was interrupted by cooling between 1945 and 1975.

(= 0.038 per cent) in 2005. The figure for 1870, at the end of the Little Ice Age, has been estimated at 290 ppm – although there has been some debate about its accuracy.[35]

Global cooling: fear of a new ice age

The evolution of temperatures did not, however, correspond to the forecasts. According to greenhouse theory it should have become warmer in the 1960s, but the opposite was the case. It is true that temperatures rose by approximately 0.6 degrees Celsius between their 1880 nadir and 1940, but for some reason the process then went into reverse. As temperatures continued to fall after 1940, no special attention was paid to the early calculations and theories of global warming.[36] For suddenly the world faced a process with which no one had reckoned, and which seemed far more dangerous: a *global cooling* lasting several decades.[37]

In the early 1960s people thought of looming ice ages; the idea of dramatic warming appeared to fly in the face of all measurements and all appearances. The starting point was the investigations of climate expert J. Murray Mitchell of the US Weather Bureau, who compared climate data with the results of recent atom bomb tests and then with those of volcanic eruptions. The dust hurled upward by atom bombs remained in the same hemisphere, whereas volcanic ash in the stratosphere could cause worldwide cooling for a number of years. But it had not prevented the warming during the first few decades of the twentieth century, nor did it explain the cooler trend in the middle of the century. That decline in global average temperatures lay outside the range of random variation. What Mitchell had in mind was the recently discovered fluctuation of the Late Dryas, some 12,000 years ago, when the climate cooled in just a few years by 10 degrees Celsius. The outcome then had been a thousand-year cold period. Was a similar freeze just around the corner?[38]

In the 1960s climatologists were obsessed with the idea of an imminent ice age. There were other reasons for this than recent environmental research. In a context of glacier growth, glaciology had broken free to become a separate subdiscipline. The ice core drilling method, which had been developed at the polar caps, showed not only that ice ages and warm ages cancelled each other out in geological time, but also that, within a given ice age, periods of greater and lesser cold – 'glacial' and 'interglacial' periods – alternated with each other. The warm period of the Holocene, going back ten thousand years, seemed to have lasted an inordinately long time. During the whole of the Quaternary, warm phases seemed always to have lasted no more than ten thousand years, whereas the remaining ninety thousand years of the Milankovic cycle had been marked by conditions of greater or lesser cold. And for two decades average temperatures on earth had been falling, as a worldwide network of measuring stations had reliably demonstrated. Did this not make it likely that the warm times were over and that the world was plunging into a new cold period?

In 1972 a group of leading glaciologists met at Brown University to discuss the looming end of the interglacial and the arrival of a new ice age. A large majority of the experts shared the view that interglacials were usually short and came to an abrupt end. They also agreed, with reference to the Milanković cycles, that the 'natural end of our warm period is undoubtedly imminent'. The most important evidence for this at the time was the cooling of the especially sensitive polar regions. All the leading climatologists concluded: 'The present *global cooling*, which reversed the warm trend of the 1940s, is still under way.' The Holocene warm period was coming to an end, unless humanity succeeded in stopping the process of global cooling.[39]

The search for anthropogenic causes of cooling

The attempt to find an explanation for global cooling did not rest content with natural causes but also searched for 'anthropogenic' factors. First in line was the atmospheric pollution resulting from industrialization and the use of combustion engines in private transport. The production of cars and lorries had assumed vast proportions in the West, where a motor vehicle soon became part of a normal household's possessions. The Soviet Union and other countries of the former eastern bloc had also gone far down the road of industrialization, and even some countries of the so-called Third World – India, China, Brazil – were heading in the same direction. In the 1960s coal and oil were being burned on a previously unheard-of scale, and raw waste gases were being discharged into the environment.[40]

In the view of some researchers, the global cooling was essentially caused by human beings: population growth, rapid urbanization and industrialization now had as much impact as 'natural processes' on the climate. The cooling was blamed on a filter effect, which no longer allowed enough sunlight to reach the earth's surface: this was called 'global dimming'. Atmospheric 'turbidity' outweighed the influence of carbon dioxide emissions, themselves also due to human activity. It was conceded that all manner of natural processes, such as desert storms or bush fires, also produced dust in the atmosphere. But the effects of big cities and heavy industry, or of car and aircraft exhaust, were much greater. Air travel, in particular, had contributed to a measurable increase in cloud cover. The piling up of anthropogenic causes thus pushed 'natural' causes well into the background.[41]

The rise in turbidity as a result of fog, cloud and smog reduced solar radiation and brought about global cooling of 0.3 degrees Celsius between the 1940s and 1970. In 1970 Mitchell – who was well aware of 'natural' climate changes in the course of the earth's history – identified human activities as the main cause of temperature fluctuation in the previous few decades. Until the 1940s, he argued, greenhouse gases such as carbon dioxide had been responsible for warming, but then atmospheric pollution had counterbalanced the greenhouse gases and produced a cooling effect. Mitchell, however, questioned whether turbidity could have been responsible for a cooling of 0.3 degrees Celsius within just twenty years, and referred instead to volcanic eruptions. To the surprise of many, he therefore predicted further warming for the future.[42]

Climatology as political futurology

With the introduction of (electronic) data processing, attempts were made to predict not only the weather but also climatic developments.[43] Since the first moon landing in 1969, the Nimbus III satellite had made global temperature measurement a possibility. But the working out of complex climate models with numerous variables only began around 1970. In 1971 leading scientists warned of the dangers of global climate change and called for the promotion of organized research. At roughly the same time, sediment drill cores and ice cores showed with increasing precision that there had been rapid climate change in the history of the planet. From 1972 to 1974, increased periods of drought and other 'anomalies' drew people's attention to the climate and the preoccupations of climatologists, with the result that concern over an impending ice age once again outweighed fears of global warming. The First Conference on the Human Environment, held in Stockholm in 1972, responded to irritation over climatic trends in the 1960s – and to the discussions of global cooling – by founding the United Nations Environmental Programme (UNEP), which was supposed to include a Global Environment Monitoring System (GEMS) to observe the effects of greenhouse gases and radioactivity on the weather, human health and animal and plant life.

Politicians in the USA and elsewhere became more attentive to the climate as a result of extreme events in various parts of the world that had led to crop damage, famine and political turbulence. The droughts of the 1970s caused not only widespread starvation in the Sahel region but also the political revolution in Ethiopia, in which a group of Marxist officers overthrew the venerable Christian imperial family. Since this meant a shift in the cold war contest between the Soviet Union and the United States, great significance was attached to the event. In a speech to the United Nations on 15 April 1974, the US Secretary of State, Henry Kissinger, pushed for increased research to meet the threat of climate change.

In that same year, a rapidly constituted Ad Hoc Panel on the Present Interglacial came to what now seem staggering conclusions: the natural climate was cooling by 0.15 degrees Celsius per annum and would thus reach 0 Celsius by the year 2015; two or three decades of slight warming would follow, at a peak rate of 0.08 degrees Celsius per decade around the year 2030; then there would be little change for another hundred years, before temperatures fell again. This absurd prognosis shows the difficulties from which climate forecasting still suffers, as its results depend on the underlying expectations and preconceptions, the operational variables and the data input. The prognosis of 1974 was utterly unsatisfactory in both methodology and content.

In 1978 the US Congress adopted a national climate programme, and the United States pressed for, and obtained, broad international research cooperation on the climate for the 1980–2000 period. Already by then climatologists credited themselves that, despite their faulty prognoses, they managed to do some good 'by undermining complacency and alerting the world community to what can happen'.[44]

For all the complacency, we should not fail to mention the practical measures that were discussed as a result of the 'certain' prediction of global cooling. Immediate danger seemed to lie ahead. If the climate was going to plunge the world into a great crisis, did we not have a duty to act quickly with the technological means at our disposal? Plans were developed to regulate the global climate by building a dam that would close off the Bering Strait between Russia and Alaska. John F. Kennedy (1917–63, r. 1960–3) seemed open to such measures during the election campaign of 1960. The Bering Dam project was seriously discussed during the presidency of Richard M. Nixon (1913–94, r. 1969–74), and it came up at a summit between his successor Gerald Ford (1913–2006, r. 1974–7) and Leonid Brezhnev (1906–82) in November 1974 in Vladivostok. The whole idea was one of the more innocuous proposals floated in those years to combat 'global cooling'.

Other suggestions were to cover the polar caps with black foil to reduce the albedo effect, or – especially original from today's viewpoint – to strengthen the greenhouse effect by increasing CO_2 production. There was also discussion of introducing metal dust into the atmosphere, building a concrete dam between Norway and Greenland, sending giant mirrors into earth orbit to function as 'extra suns', and creating an artificial 'ring of Saturn' around the earth, made of potassium dust. The military establishment felt equally moved to contribute. Its ideas included blowing up undersea mountains with atom bombs southwest of the Faroe Islands, in order to extend warm ocean currents into the Arctic; heating Greenland with the help of nuclear reactors; and melting the polar ice with hydrogen bombs. These plans sound straight out of the world of Dr Strangelove. Even at the time they were considered too problematic for public discussion, but they were kept in reserve in case the cooling became more intense.[45]

A new approach to warming and the dispute over its causes

Meanwhile Syukuro Manabe and R. T. Wetherald calculated that a twofold increase of atmospheric CO_2 levels would cause a temperature rise of several degrees Celsius.[46] In 1970 Manabe went so far with his sums that, on the basis of rising CO_2 emissions, he thought he could predict a 0.6 degree warming by the end of the century.[47] The growth of air travel, with its related exhaust gases and haze effects, led in 1975 to research into its environmental spin-off and the wider issue of trace gases in the atmosphere. This work revealed the dangers of CFCs (chlorofluorocarbons) and their possible contribution to the greenhouse effect. Growing ecological awareness also focused attention on deforestation and other human interventions in the ecosystem. While many researchers still considered how the atmosphere could be artificially heated, others were asking whether we were not on the brink of a new global warming.[48]

Around 1977 a new consensus began to form among scientists that global warming was indeed the greater threat. In 1978 a National Climate Program Act hugely increased research funding, and in the same year Stephen H. Schneider founded a new scholarly journal, *Climatic Change*.[49] The National Academy of Sciences found it credible in 1979 that a doubling of atmospheric CO_2 levels would result in global warming of 1.5 to 4.5 degrees Celsius.

The election of Ronald Reagan set the climate research community at odds with the US government, which took a sceptical attitude to the predictions of global warming. Climatologists resisted political pressure, however, and identified greenhouse gases, above all carbon dioxide, as the cause of warming,[50] especially when 1981 proved to be the hottest year since records began and sharp temperature rises occurred in Greenland. Global warming was now seen as an incontrovertible trend.[51] Some governments, especially in the USA, Australia and Britain, began to plan responses to climate change – for example, measures relating to water conservation, ship transport and agriculture.[52]

Since the 1980s the media have been shocking the public with one doomsday scenario after another, but they have also presented serious information about the signs of impending climate change, such as the retreat of glaciers in mountainous and polar regions or changes in the flora and fauna. The media raised the stakes when several years of drought and record heat (for example, the dry years in the British Isles between 1988 and 1992)[53] seemed to indicate that something was wrong with the climate. Statements made to the US Congress by Stephen H. Schneider of the National Center for Atmospheric Research played a key role in alerting the public that global warming had begun. Since then, respected scientists have built the theory of anthropogenic global warming into their text-books.[54] The debate on the causes of global warming has been conducted more publicly than almost any other disputed issue in the sciences.[55] And, of course, public interest and greater investment in climate change research have led to an unparalleled widening of the scientific focus.[56]

Reactions to Climate Change

Climate science and global warming on the international agenda

The establishment of the GEMS monitoring system was not without consequences. Its worldwide data showed that the burning of fossil fuels, large-scale deforestation and changes in land use were year by year raising CO_2 levels in the atmosphere. In 1988 the first World Climate Conference, held in Toronto, appealed for discussion of these results and issued a stark warning of the dangers of global warming due to higher atmospheric levels of CO_2 and other greenhouse gases. Its closing declaration urged taxes to counter these effects immediately. In 1988, under UN auspices, the Intergovernmental Panel on Climate Change (IPCC) was set up in Geneva to coordinate climate research and protection around the world.[57] Its mission is not actually to promote research but to issue every five years a comprehensive, non-partisan and transparent report on the state of knowledge concerning the climate and the consequences of climate change, with a view to reaching a consensus. The first such IPCC Report, published in 1990, received great attention.[58]

At first, however, not all scientists shared the view that a new long-term trend had begun. After all, there were still glaciers that did not melt, and some were even growing as a result of increased precipitation. Glacier research thus became a political issue. In the United States, the tensions grew more serious between climatologists and liberal public opinion and the administration of the forty-first president, George Herbert Walker Bush (b. 1924, r. 1989–93). A new US Environmental Protection Agency, closer to the White House, was set up as an alternative voice to that of the long-established US Academy of Sciences. When the forty-second president, Bill Clinton (b. 1946, r. 1993–2001), took office, the situation improved for conservationists and climatologists, especially as a leading politician, Vice-President Al Gore (b. 1948), took a number of policy initiatives on the environment.[59] In the mid-1990s opinions were still divided over the fact or the causes of global warming, with apocalyptics and sceptics bitterly opposed to each other.[60] The first IPCC Report in 1990 created the basis for the United Nations Conference on Environment and Development, also known as the Earth Summit, held in Rio de Janeiro in 1992 and attended by ten thousand delegates from 178 countries. One of its five documents was the United Nations Framework Convention on Climatic Change, which came into force on 21 March 1994 and marked a turning point in climate policy. The signatories declared their willingness to protect the earth's ecosystem for the use of present and future generations, so that food supply would not be threatened and species adaptation to climate change as well as sustainable economic development would continue to be possible. The production of greenhouse gases was to be restricted or reduced. The aim was to stabilize greenhouse gases at 1990 levels by the end of the decade.

After the Framework Convention came into force, the contracting states agreed at their first follow-up conference (Conference of Parties, COP 1) in Berlin in 1995 that the Rio commitments were not sufficient and that the industrial countries bore the main responsibility for climate protection. A commission was appointed to develop within two years a set of concrete tasks and deadlines for the reduction of greenhouse gas emissions. In 1996 a second follow-up conference (COP 2) in Geneva discussed the second IPCC Report, issued in 1995.[61] Government representatives recognized for the first time 'evidence of human influence on the global climate'. The greatest step so far to make greenhouse gas emission limits binding under international law came at the third follow-up conference (COP 3), held in Kyoto in December 1997.

Countries that signed the Kyoto Protocol undertook to lower emissions of carbon dioxide and other greenhouse gases by 5.2 per cent in comparison with 1990 levels, and to achieve this by 2008–2012. Some participants, such as China and India, did not accept any restrictions because of their development lag, but the European Union committed itself to a reduction of 8 per cent, the United States to 7 per cent and Japan and Canada to 6 per cent. On the grounds that the greenhouse effect is global, industrial countries that exceeded their limits could purchase pollution rights from developing countries that were not using up their quotas. The Kyoto Protocol was supposed to come into force as soon as it was ratified by 55 states responsible for at least 55 per cent of anthropogenic carbon dioxide emissions in 1990.[62] However, the ratification process took considerably longer than expected.

The IPCC Report of 2001

In 2001 the Intergovernmental Panel on Climate Change (IPCC) published its third commissioned report on the state of climate research (Third Assessment Report); it was based on the work of 426 experts, and its results had been checked at two stages by a further 440 advisers and supervised by 33 editors. There is no other scientific publication on which so many experts have worked together. The IPCC membership consisted of scientists and government representatives, including from the United States (the largest oil consumer) and Australia, which had not signed the Kyoto Protocol. Representatives from Saudi Arabia (largest oil producer) and China (largest coal consumer) were able to insert their reservations about certain formulations – not an unimportant point, since IPCC reports require unanimity before they are issued. It is precisely this balance of opinions which gives IPCC reports their high reputation.[63]

The 2001 report was preceded by a Special Report on Emission Scenarios, which presented forty different economically plausible models for the period up to the end of the twenty-first century. The most optimistic of these was based on a negligible rise in CO_2 emissions, followed by a gradual decline to a fraction of today's levels; the most pessimistic assumed a fourfold increase in CO_2 emissions by the year 2100. The projections point to a rise in atmospheric CO_2 between 540 and 970 ppm by the target date. But, with negative ocean and biosphere feedback, the figure could be much higher. As to temperatures, the models imply a global mean rise of 1.4 to 5.8 degrees Celsius.[64] Many climatologists thought even these projections too optimistic. Representatives from the Potsdam Institute of Climate Impact Research considered warming of less than 2 degrees completely implausible, but 8 to 9 degrees quite possible. The EU's target of 'only' 2 degrees, at the lower end of IPCC projections, thus appeared to them unrealistic.[65]

Even warming of 2 degrees, however, would be far greater than temperature variations experienced over the last few centuries. Global

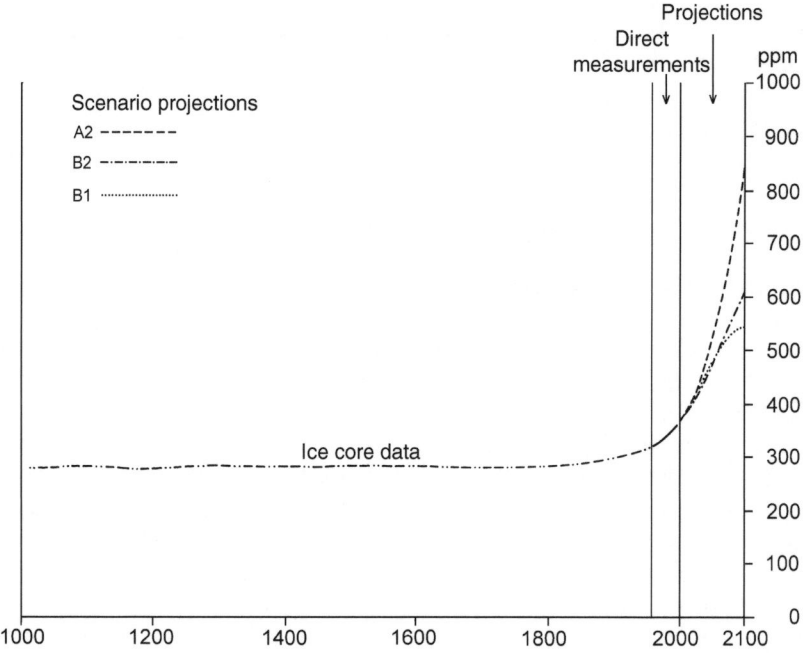

5.6 The hockey stick in the IPCC Report of 2001 has something puzzling. If the CO_2 component of the atmosphere determines temperatures, and pre-industrial CO_2 was a constant 280 ppm, what accounts for temperature variations over the past thousand years? Is the hypothesis false? Are the measurements wrong? Or were the statistics simply doctored to dramatize the climatic change?

warming in the whole of the twentieth century reached 0.6 degrees Celsius. Mean warming during the Roman optimum or the medieval warm period probably totalled no more than 1 to 2 degrees, and mean cooling during the Little Ice Age also probably did not exceed 1 to 2 degrees. We have seen how large were the effects of such small temperature variations. Global warming of 6 degrees Celsius would have effects of a magnitude that we can scarcely imagine. Consequently, the third IPCC Report, issued in 2001, paints a more dramatic picture than those contained in the first two. It lays greater stress both on the warming trend and on the guilt of environmental sinners. It explicitly confirms the human contribution to global warming and plays down the role of natural variations.[66]

Signs of a new awareness

Not least because of the third IPCC Report, the Kyoto process gathered momentum again before the follow-up conference in 2001 in Marrakesh, even though the United States announced that it was withdrawing from the process. The victor in the presidential elections had not been the conservationist Gore but the representative of the oil and weapons industries, George W. Bush. It was now clear that a reduction of greenhouse gas emissions would not be one of the priorities of US policy. After the attack on the World Trade Center in September 2001, the fight against international terrorism became its number one concern. Nevertheless the Kyoto process now made rapid advances: Iceland's ratification on 23 May 2002 gave the Protocol its crucial fifty-fifth signature, and, after lengthy negotiations, the Russian president Vladimir Putin finally took the plunge on 18 November 2004. The Kyoto Protocol came into force three months later, on 16 February 2005, signed by 141 states representing 85 per cent of the world's population. The Montreal conference of contracting parties (COP 11), held in November–December 2005 with roughly ten thousand delegates and observers from 188 countries, was the first after the Protocol came into force and thus became the first Meeting of the Parties of the Protocol (MOP 1). Negotiations began there on climate protection measures for the period after 2012.

The negotiations on climate protection are among the most interesting processes in a world which, at the political level, has been learning to conceive itself more and more strongly as a global society. The only other examples of such a wide-ranging international agreement are the negotiations that led to the United Nations or the EU integration process. In the end, however, the world's largest emitter of carbon dioxide, the United States, refused to sign the document, and doubts have therefore been expressed about its usefulness.[67] Other countries such as Australia, China, Saudi Arabia and the Group of Developing Countries have for various reasons displayed a negative attitude to the updating of the Kyoto Protocol after 2012. Nevertheless, the Montreal conference cleared the way for negotiations on precisely that goal. The conference chairman, Canada's environment minister Stéphane Dion, even spoke of a future-oriented Montreal Action Plan (the MAP of the MOP). And, at the end of the conference, although the United States had already announced its withdrawal, the leader of the US delegation showed a willingness to engage in strategic dialogue. This was not least a result of Hurricane Katrina, which in August 2005 devastated New Orleans and parts of the American Gulf coast and caused President Bush's poll ratings to nosedive.[68]

The Fourth IPCC Report (2007)

On Saturday, 3 February 2007, global warming finally reached the front pages of serious newspapers. The political sections, columnists and cultural pages also reported on it in detail, in the last case probably to explain its importance for culture. The previous afternoon, in fact, the latest conference on the global climate had ended in Paris with an official presentation of the fourth IPCC report on international research into climate change. The current chair of the IPCC, the Indian Rachendra Pachauri (b. 1940) elected in 2002, spoke at a press conference on the results of the 500-strong Working Group I, which contained the scientific underpinnings of the whole report.[69]

In comparison with the 2001 Report, the 2007 'summary for policy-makers' – which can be downloaded from the internet – contained some interesting shifts of emphasis. Climate scientists no longer left much room for doubt about the anthropogenic element in global warming: they thought it 90 per cent certain, though not, as the press claimed, 'indisputable', since there is seldom absolute certainty in the sciences. Representatives of 113 participating governments, who had been able to influence a report requiring unanimity, agreed with this assessment. Unlike in earlier reports, the formulations were not softened or watered down in the draft – and the final version made some of them even sharper.[70] In his summing up, Achim Steiner from Brazil (b. 1961) stated that those in positions of political or economic responsibility were now called upon to make a decisive response. The current director of the United Nations Environmental Programme (UNEP) thought that, because of the final clarification of the 'anthropogenic warming' issue, 2 February 2007 would go down in the history books.[71]

Climatologists stood firm that the main cause of climate change was 'anthropogenic gases': primarily CO_2, then CH_4, N_2O and tropospheric ozone. Atmospheric emissions of the most important greenhouse gas, carbon dioxide, had increased considerably even since the 1990s: from 23.5bn tonnes to nearly 26.4bn tonnes per annum. In addition, indirect emissions from forest clearance and changed land use accounted for another 1.8bn tonnes to 9.9bn tonnes. Natural causes of warming – such as slightly increased solar activity – scarcely entered into the equation. On the other hand, anthropogenic pollution by the fine particles known as 'aerosols' was a factor in the

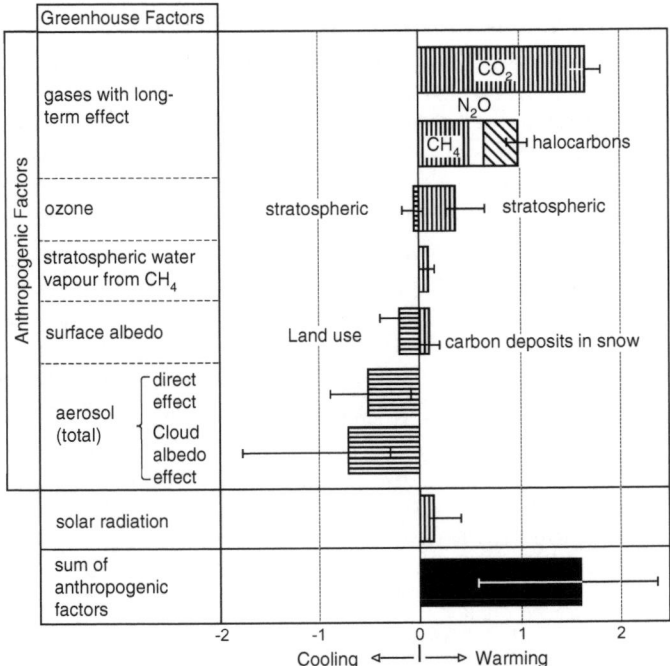

5.7 Anthropogenic factors have a double-edged effect on the climate. According to the IPCC Report of 2007, the factors that promote planetary warming have been uppermost since the 1970s.

increased cloud formation, all of which intensified the albedo effect; but for this 'anthropogenic cooling effect' the warming would have been even greater. In the IPCC view, the total 'anthropogenic warming effect' far outweighs natural influences on contemporary climate change. Interestingly, the IPCC Report refers at this point to some data on palaeoclimatic history. Basing themselves on ice core samples, climatologists now think they know the composition of the air over the past 800,000 years, and hence also the share of trace gases, including the greenhouse gases CO_2, CH_4 and N_2O. If ice core readings for the past 10,000 years (the Holocene warm period) are combined with instrumental measurements for the past hundred years, the picture is one of relatively constant levels: approximately 275 ppm for CO_2, for example (but see figure 1.7).

Over the past two hundred years, however, but especially over the past fifty, what we see is a rapid increase in these levels up to the present day or, in the diagram, up to 2005. The current concentration of 359 ppm of CO_2 is already far in excess of the highest known level for the past 650,000 years. According to theory, which links temperature rises to higher concentrations of greenhouse gases, we should therefore expect a major increase in global mean temperatures. In comparison with the average for the 1901–50 period, the global mean temperature has risen by roughly 0.6 degrees Celsius – a little more on land than on ocean surfaces. But in comparison with the average for 1850–99 – the last years of the Little Ice Age – the increase on land surfaces is significantly greater: ±0.76 degrees. This is the figure usually quoted in the press, because it makes the global warming appear more dramatic. According to the Report, actual temperature measurements for the period from 1906 to 2005 in all continents are highly consistent with the calculations in the model. In all cases it turns out that the temperature increase has been greater since the 1970s, and especially in the past twenty years. The accelerated warming has also affected deeper waters, being discernible as far as three thousand metres under the sea.

The warming has gone together with a measurable rise in sea level. Over the twentieth century this increased by approximately 1.5 millimetres a year – or a total of 15 centimetres. But, within this fairly undramatic framework, there was a definite acceleration in the 1990s, and the level of the sea is currently rising by 3.1 millimetres per annum: a little more than 3 centimetres every ten years. Despite the problems of satellite-based calculation – the determination of average values is methodologically difficult because of the earth's potato shape and differential gravity – there is no disputing the fact of the increase. Less clear is what has been causing it. The warming itself extends the ocean surface and induces higher water levels. But the degree to which this is due to the melting of mountain glaciers and polar caps can be only approximately calculated. There is also argument about the likely further rise in sea levels over the coming decades. The IPCC report of 2007 has revised downward the figures in the 2001 report: a rise of 18 to 59 centimetres by the year 2100, instead of the earlier forecast of up to 88 centimetres. One reason for this is that no great melting has been observed or is expected at the South Pole, since temperatures there are on the whole too low. Owing to increased precipitation, it is even expected that the Antarctic ice sheet will grow. In terms of mean sea levels, we are therefore talking of something like a rise of 35 centimetres – which would make groundless any fear that the Pacific isles, Florida or Bangladesh will disappear over the next three generations.

The IPCC commissioned forty models for the twenty-first century, based on various scenarios for the development of greenhouse gas emissions. These present the expected warming of the earth in relation to average temperatures for the 1980–99 period. The model calculation families B1, A1B and B2 are the ones with the highest degree of probability. The *A1 model* assumes fast growth of the world economy and world population growth until the middle of the twenty-first century (followed by a period of decline), as well as the rapid introduction of new and more efficient technologies. The A1 group comprises three variants, which either continue to rely on fossil energies (A1FI) or use non-fossil energies (A1T) or a mix of all possible energy sources (A1B). *A2 scenarios* describe a completely different world. They assume a continuation of world population growth and heterogenous regional developments, with very uneven and altogether slower economic growth. They will be called for short the 'keep on as before' scenarios. *B1 models*, by contrast, assume economic convergence, with demographic trends as in A2 but with a rapid structural shift to a service and information economy and the introduction of cleaner technologies. This is the scenario of the bio-industrial technological revolution. *B2 models* assume a constant but slower growth of world population, medium economic growth-rates, slower technological change and a growing environmental awareness that achieves a breakthrough in certain regions.

All the models are considered equally likely. Globally averaged, they translate into warming of 1.5 to 4 degrees Celsius by the year 2100 – which is evidently less than apocalyptic forecasters expect. The effects of even two degrees may still be considerable, of course, as we have seen from the precedent of the Little Ice Age. In all the models, the warming at first mainly affects the northern hemisphere, since the Antarctic, with its constantly low temperatures, will probably not suffer any major ice loss before 2030; the ice cover may even become more extensive there as a result of increased precipitation. Only in the second half of the century will temperatures also rise in the southern hemisphere and even Antarctica, by roughly two degrees in model B1 and four degrees in B2. With temperatures below –30 degrees, however, there will still be no danger of a thaw. The situation looks quite different at the North Pole. Here the rise in average annual temperatures will already be more than two degrees by the year 2030 and, according to the model calculation, will be of the order of 6–9 degrees by the end of the twenty-first century. The IPCC therefore calculates that the sea ice will melt in summer, and in the longer term perhaps even Greenland's inland glacier will melt away.

In temperate regions, of course, the picture looks different again. Warming of only one degree is expected in Western Europe by the year 2030, and 2–4 degrees by 2099. The warming will be greater for North America and North Asia, and in the B2 model also for South America, southern Africa and Australia.

The great migration has begun

We know the consequences of previous warmings: glaciers retreat, the tree line rises, vegetation creeps closer to the poles, and animals follow in its path: insects, amphibians, birds, fish and mammals. The flowering of trees and blossom time, the nesting of native birds and the arrival or departure of migrants no longer take place at the same time as before. The sea grows warmer, ocean levels rise, disastrous floods become more frequent, and patterns of habitation change. All this happened in the great Holocene warming ten thousand years ago, as it did in lesser warmings such as the one at the beginning of the Atlantic period or during the short Roman optimum. The current warming goes back only to the 1890s or – if we start from the end of the cooling in the mid-twentieth century – only to the 1970s. It has thus lasted for little more than a generation and led to temperatures only 0.6 of a degree higher. Nevertheless, we can already see some effects of the global warming or what we might call the 'modern optimum'. A recent synoptic study of the behaviour of 677 animal species concluded that 62 per cent of the species had shown signs of climatic warming in the last two decades and 27 per cent had displayed no significant change, while 9 per cent had reacted as if spring had been delayed.[72]

In a densely populated world, in which pieces of land are fenced off and animals live in protected areas, migratory movements are no longer as easy as they were in the Palaeolithic. But unfettered movement is still possible across the skies and oceans, and the inhabitants of those elements follow the food supply. If reforestation occurs in northern Alaska or Siberia, insects and forest animals start to show up there again. The anopheles mosquito is again finding an abode north of the Alps and spreading into North America. Butterflies are heading up towards the pole.[73] Only rarely do they attract as much attention as the aptly named hummingbird hawkmoth, whose unmistakable proboscis works away at garden and balcony flowers far away from its Mediterranean homeland. It is capable of flying up to two thousand kilometres and crossing the Alps without too much difficulty. For some years now it has been starting to settle north of

200

the Alps. Many specimens spend the winter there, and in March lay the eggs from which their offspring will hatch in mid-June.[74]

Species diversity has also been increasing in mountainous regions. This is a disadvantage for species that have specialized in the adverse conditions of life at great heights or in the far north. The habitat of polar bears is shrinking in favour of forest bears. Migratory movements can also be detected in the oceans. For example, a shoal of tropical sunfish apparently followed a shoal of tasty jellyfish to the waters off Cornwall. Species of tropical fish are venturing into the Mediterranean or the North Atlantic; cod are migrating further north and can be caught in places where this last happened in the Middle Ages.[75] The geographical limits for cultivated plants are again moving north. At the beginning of the twenty-first century, new areas opened up for quality wines in Belgium and the German state of Mecklenburg. But this is not as far north as in the Middle Ages, nor has Greenland seen the return of cereal or livestock farming. Viking graves still lie there under permafrost, but it seems only a question of time before they can be easily dug up again.

Migration is not possible everywhere in the modern world. Wildlife reserves might turn out to be traps, since their inhabitants are not capable of relocating through their own devices. The same is true of many plants, whose seeds are unable to survive outside protected areas or ecological islands; and those which live on islands face the problem that they lie only just above sea level. Admittedly it would still be an exaggeration to speak of climate refugees, as the curious story of the vanished village on the Vanuatu atoll of Tegua demonstrates. This village was tormented by an earthquake, a tsunami and a hurricane, yet each one passed through and had nothing to do with the two centimetre rise in sea level that has been recorded up to now. So, we have to be careful with this example. After the island-dwellers had cashed in their aid money from a climate fund, they even refused to move to higher land within the same atoll.[76] This is not to say, however, that there will not be major problems in the future. Were the sea level to rise by one or two metres, as apocalyptic analysts predict, low-lying countries such as Bangladesh or perhaps even the Netherlands would be in serious trouble.[77]

The likely consequences of global warming

The social and political consequences of global warming will be enormous. But they will also be highly varied for both animals and national economies: some will benefit, others will suffer. It is by no

201

means clear how the future opportunities will be distributed. Increased solar radiation may mean both greater difficulties for agriculture and greater advantages from solar energy use. It is predicted that global warming will lead to periods of drought in many regions that are known to have profited from earlier warm periods: one need think only of the well-irrigated Sahara in the Atlantic period (see figure 2.2). The way the scales tip will perhaps depend on adjustments at the level of the economy, that is, on a factor only indirectly linked to the climate. In the past, the rise and fall of civilizations was determined by such things as their innovative capacity and adjustment to market requirements. It would be too much to claim that Britain, the world's first industrial nation, had a particularly auspicious climate. But cultural traditions, political, social and religious conditions, did play a role. One of the achievements of industrial civilization is that it made people less dependent on climatic influences than at any time before in the long history of human life on earth.

Global warming will not change any of this in the foreseeable future. The large industrial countries will probably not face huge difficulties in coping with it. The United States of America, the European Union, Russia, China, Japan, Australia and Brazil will have to come to terms with increased internal migration and a restructuring of their housing markets, but it is unlikely that this will change much as far as their role in the world is concerned. Other countries will find their niches, as they have done up to now. Material wealth and cultural openness will have some influence on how they handle the need for adjustment. There will certainly be changes in the behaviour of tourists, who will look around for different places to go.[78] And there will certainly be effects on the property market and, more generally, on investors' decisions.[79]

Among those who will suffer are indigenous peoples under adaptation pressure from Western lifestyles, whose own cultures often do not provide the resources to survive in changed climatic conditions. Other losers will be socially deprived groups unable simply to relocate from their submerged little house on the North Sea coast to Majorca – or vice versa. People attached to their native land will always lament its loss. But nor should we underestimate the ability to sink roots in a different region. The European continent, with its whole painful history of expulsions and forced emigration, has learned that it is possible to make a fresh start, especially when society offers loans or other kinds of support. In future, perhaps such assistance should no longer be tied to membership in the same ethnic or linguistic group; perhaps new tasks will fall to the United Nations in the management of migration flows. In any event, the resulting problems should not be exaggerated. The population of Pacific atolls or northern Alaska is lower than that of small towns in America, Asia or Europe. The victims of drought in Africa or Asia weigh more heavily in the equation, and in the long term large migrations are expected from there as a result of desertification processes.[80]

The 'save the world alliance' and Plan B

As a well-known journalist remarked, 'global warming, the mother of all environmental crises, has triggered a political movement in the last twenty years which has so far had virtually only one orientation: Down with greenhouse gas emissions! . . . It has fought against carbon dioxide emitters at any price and with all the weapons of global scientific expertise, in a kind of travelling meta-panel whose members have moved purposefully from one conference to the next. The Climate Framework Conference and the Kyoto Protocol are already classic examples.' Among the climate experts, there has 'for years been something like a political-ecological ethos that outlaws any significant deviation from the model of emission reduction'. An effective brake on warming, however, would require much more than the Kyoto objectives to be achieved by the middle of the century – a heroic or utopian goal, but one that is out of the question under the given political conditions. 'On the other hand, adjustment measures such as those which affected countries take in response to warming, desertification and flood threats are regarded by many as secondary in importance, or even as unproductive in terms of communicating the problem. The same applies to technical solutions that go beyond energy technology. Industrial countries, according to the "politically correct" view, are the problem and therefore cannot be the solution.'[81]

Yet, despite ratification of the Kyoto Protocol, the world is still a long way from a reduction of greenhouse gases. Emissions are continuing to increase, because leading industrial countries such as the United States are not part of the Kyoto process, while developing countries such as China, India and Brazil have negotiated unlimited pollution rights. Even environmentalists from these countries believe that only the rich should pay for climate protection, whereas the underdeveloped countries should have the right to pursue catch-up industrialization. The Indian anti-pollutionist Sunita Narain argues: 'Climate change raises the issue of justice, of a fair distribution of resources.'[82] Even some industrial countries that were among the first

5.8 'Even now our CO_2 emissions are too high.' An Ark for the new Flood, but even the rescue ship emits CO_2. And instead of animals we save our lovely car.

to sign the Kyoto agreement – Spain, for instance – have not been sticking to the upper pollution limits. 'Despite everything, the "save the world alliance" is facing disaster, since the solution to the energy question is taking too long and the global release of climate-impacting carbon dioxide continues without restriction. The scientific world – or one part of it – has therefore evidently opted for Plan B.'[83]

Instead of, or in addition to, a reduction in waste gases, this Plan B envisages a technological solution to the greenhouse problem; its keyword is *geo-engineering*. One proposal along these lines, devised by the US geologists Klaus Lackner and Kurt Zenz House from Harvard, would involve the sequestration of CO_2: the greenhouse gas given off in the use of fossil fuels would be filtered in grand style and permanently disposed of in suitable ways – for example, in exhausted coal seams, in the ocean depths or even in seabed sediments. A special IPCC report seriously discusses such brakes on climate change, and they have received the backing of climate scientists.[84] The prospect of a geosequestration policy has already triggered a kind of green gold rush among geological service providers, who hope that it will bring them good business. If it proves successful, the policy would result in emission-free consumption of fossil fuels.[85]

Even more extravagant seem proposals to delay global warming by strengthening the albedo effect in the stratosphere; wafer-thin, light-reflecting particles of sulphur would beam sunlight back into space at a height of fifteen kilometres above the earth. Since Nobel prizewinner Paul Crutzen endorsed the idea, there has been serious discussion of how to introduce sulfate powder into the stratosphere with the help of balloons, rockets or cannons, as a way of reducing the greenhouse effect. According to Crutzen's calculations, sufficient artificial pollution of the stratosphere would require no more than five million tonnes of sulfate powder a year, less than a tenth of worldwide sulfate emissions, at a cost of $50 per head of the population in the industrial countries. Nor would the veil of sulphur significantly impair our experience of nature: gloriously coloured sunsets would continue, only a little paler than we are used to – not a high price to pay for keeping our familiar ambient temperature and familiar climate.[86]

As one might have expected, this technocratic venture did not win much support among those who largely blamed global warming on human influences. The same is true of other flights of fancy, such as the iron fertilization of ocean plankton to remove CO_2 from the planet's carbon cycle, or large-scale reforestation and the production of genetically modified plants to increase carbon binding, or the installation of giant mirrors in space, or the irrigation of deserts to reduce the albedo. Schellnhuber and Rahmstorf describe these technofixes as an attempt to manipulate the earth system.[87] The idea of allowing a Dr Frankenstein to experiment with the world's climate has aroused widespread concern, since a mistake here would have far-reaching consequences.[88] We should remember the fixes for supposed global cooling in the 1970s, and consider how absurd they sound today at a distance of just one generation.

— 6 —

EPILOGUE: SINS AGAINST
THE ENVIRONMENT AND
GREENHOUSE CLIMATE

Climate reading as a new religion

The American environmental activist Robert F. Kennedy Jr, nephew of the famous president JFK, wrote shortly after Hurricane Katrina swept through Mississippi that the State governor, Harley Barbour, bore some of the responsibility for what had happened.[1] In his view, this oil industry representative and member of the Bush election campaign team had pressurized the new President to go back on Bill Clinton's environmental policy, to withhold his signature from the Kyoto Protocol, and to ignore the statements of scientists about the impending climate change. A study conducted at the Massachusetts Institute of Technology, and published in the journal *Nature*, had shown that the frequency of destructive hurricanes was attributable to global warming caused by human activity. The US dependence on oil had led not only to the horrific Iraq war but also to Katrina, which gave some idea of the climate chaos we would be leaving to our children. In conclusion, Kennedy invoked a Christian fundamentalist as the star witness: 'In 1998, Republican icon Pat Robertson warned that hurricanes were likely to hit communities that offended God. Perhaps it was Barbour's memo that caused Katrina, at the last moment, to spare New Orleans and save its worst flailings for the Mississippi coast.'[2] The trope of divine retribution for human sins has not yet faded into oblivion, even today in the twenty-first century.

'Sins against the environment' or 'ecosin', which so reminds us of the sin economics of the late Middle Ages and the early modern period, is meant not as a scientific term but as a religious metaphor. It occurs even in publications by noted scientists – for example, geo-scientist Richard B. Alley's contribution on past climate change.

Although there have always been sudden swings – Alley mentions the abrupt cooling of the Younger Dryas – the journal editor adds in an explanatory box: 'Climate experts are only just beginning to understand what triggers such swings. But it is fairly certain that ecosins such as the massive emission of greenhouse gases increase the risk of a sudden, long-lasting climate change.'[3] There is no precise definition of what is meant by 'ecosin', even in publications that have the term in their title.[4] It is clear that we are here on the ground of religion rather than science. In a theological perspective, a sin is an offence against God's command and therefore deserves to be punished. In earlier societies it was the task of priests to point out violations of divine law. Today this role seems to have been taken over by climate scientists.

The legend of a lost natural balance

Alongside the sinfulness of God's creatures, another intellectual construct features prominently in the argument: that of a 'natural balance' or sometimes 'climatic equilibrium'. For example, James E. Hansen, director of NASA's Goddard Institute for Space Studies and professor at the Earth Institute of Columbia University (New York), writes that global warming throws the earth into 'energy imbalance'.[5] But Hansen forgets to mention exactly when he thinks the climate was in 'balance'. Does he, like the late Ellsworth Huntington, mean the kind of temperate climate that exists on the East Coast of the United States, which alone permits complex civilizations to prosper? Even during the Holocene the climate has never been constant. Over the last five billion years – since the formation of the earth – it has always been changing, and it will continue to do so in the future. It might therefore be said that it is always in balance, because all climatic factors interact with one another and the result can by definition only be a state of balance.

The image of a lost balance takes us onto the ground of medical metaphors, which enjoy growing popularity among climatologists and journalists.[6] As in Galen's theory of illness in antiquity, which presented health as a state of balance among four bodily humours, an imbalance in nature is supposed to have made it sick. Two physicists, who in different capacities have been coordinating their work with the IPCC, claim that the recent warming is at most comparable to a fever: global temperature and precipitation increase, as do sea levels more slowly, and extreme weather situations become more frequent. Their otherwise finely nuanced argument is summed up in

6.1 'The effects of climate change are more and more disastrous!' Nightmare of the apocalypse theorists. The climate becomes a topic of party chatter, and no one likes to get worked up about it.

the image of illness: 'The IPCC's most important point is that the climate change disease will persist and have both useful and harmful effects on ecological and socio-economic systems.'[7]

So, the earth is sick and needs a doctor. It has the climate disease, marked by an imbalance of humours and high fever – or a slightly raised temperature and a medical expectation of high fever. That this metaphor borders on nonsense also occurred to the editor of the *Erde im Treibhaus* dossier. For he writes apologetically in a special comment: 'The understanding of scientific connections lives off images, similes, metaphors. . . . But images also harbour dangers. Metaphors may be distorted and inchoate, may be overdrawn, may lead to wrong conclusions and involve illegitimate simplification. . . . So, care is needed with comparisons. This is also true of the "climate illness" of planet earth. Although the image may make things easier to understand, the reality is undoubtedly more complex than the representation.'[8]

The 'Anthropocene'

In the year 2000, the Dutch Nobel prizewinner Paul Crutzen (b. 1933), former director of the Max Planck Institute in Mainz known for his research on the ozone hole and the chemistry of the atmosphere, claimed that humans have impacted so strongly on the climate that we can no longer speak of a 'natural' climatic period. Since the beginning of industrialization – which Crutzen dates to the invention of the steam engine in 1784 – man's artificial production of trace gases (especially CO_2) has so changed the earth's atmosphere that we have to assume we are at the start of a new age. The Holocene is over, and a new age shaped by humans has begun: the Anthropocene.[9] Crutzen refers to the point that interglacials have seldom lasted longer than ten thousand years. His concept of the Anthropocene implies that human assaults have upset this 'natural' rhythm, and that now we are heading for further warming instead of a period of cooling.

The idea of a manmade climate was soon being embroidered in a way that surprised its originator. William F. Ruddiman (University of Virginia, Charlottesville) advanced the thesis that the natural cycle was disturbed much earlier, with the advent of agriculture. From the evidence of ice cores, he discovered that the methane composition of the atmosphere changed around that time and had an early impact on the climate. In this reading, the methane composition was consistent with Milanković's predicted intensity of solar radiation until about eight thousand years ago. Methane (CH_4) is produced in large quantities by marshes and swamps, so that more of the gas is generated in warm moist ages than in dry cold ones. After the end of the Holocene thermal maximum, one would expect there to have been a decline in methane composition, gradual at first but then accelerating about five thousand years ago. Instead, however, we find in the ice cores only a superficial dip followed by another increase. Similarly, Ruddiman thinks he has discovered irregularities in the CO_2 composition of the atmosphere, which he takes as proof that, since the Mesolithic, humans have wrested control over the climate from nature.

Widespread fire-clearance of forests in Eurasia and America, and above all the beginning of rice cultivation in East Asia and the systematic raising of livestock, produced huge quantities of trace gases and loaded the atmosphere with them. The building of weirs and dams to regulate the water supply created giant artificial swamps at certain times of the year. Ruddiman calculates that, over the thousands of years up to the high Middle Ages, anthropogenic effects had an impact on the climate as high as 2 degrees Celsius. This was partly 'masked' by a global cooling trend – although this point is hard either to prove or to disprove. In short, in Ruddiman's account, the Holocene is substantially identical with the Anthropocene.[10]

At first sight, the thesis of significant human influence on the climate since the beginning of the Holocene appears absurd, since the world population ten thousand years ago was minute in comparison with the present day. For the initial phase it has been estimated at no more than 7.5 million, and over the next eight thousand years it grew to 300 million.[11] Changes in the landscape, which during that time were effected with simple technological means, nevertheless had a huge impact on the surface structure of the planet. Clearances and cultivation already banished most of the European forest in the Bronze Age; the same happened in the Middle East and North Africa, the Far East and parts of North America (where indigenous farmers dried swamps on a grand scale – a point that would seem to contradict the methane thesis). The radical alteration of the earth's surface was a great climatic experiment, since it changed the albedo and the composition of the atmosphere.

In response to Ruddiman, Crutzen attacked his suggestions and sharpened his own arguments. In his view, it was unimportant when the Anthropocene began – 8000 or 5000 BC – since humanity has undoubtedly had more influence on the environment since the Neolithic Revolution than at any time before. It is also clear, however, that this impact has increased by leaps and bounds in the two hundred years since the beginning of the Industrial Revolution – not immediately, but persistently. The world population has meanwhile increased tenfold to 6 billion, and with the estimated 1.4 billion cattle there is probably more livestock on earth than at any time before. Moreover, Ruddiman argued, there was another sharp break around the year 1950. Since then, human activity has changed nature on a scale that was previously inconceivable. It is estimated that, by the end of the twentieth century, deforestation, agriculture, stock-farming and construction had transformed 30 to 50 per cent of the planet's land surface. Until mid-century humans influenced the environment and the climate; from then until now human influence has dominated every aspect of the earth system, from the atmosphere to the land and ocean and coastal regions. This is true also of the climate. Agriculture, nitrogenous fertilizers and fossil fuels pumped more greenhouse gas into the atmosphere than natural processes could possibly have done. A sequential model of the Anthropocene might show the Neolithic Revolution as stage one and the Industrial Revolution as stage two; but a third stage would begin around 1950, with 'very significant acceleration' of many influences on the earth system. A fourth stage might begin in the twenty-first century, hopefully marked not by further pillage of natural resources and environmental pollution, but by responsible intercourse with the earth system, population control and conscious environmental management.[12]

Ruddiman's expansion of the Anthropocene era shifted attention from vexing contemporary phenomena and uncertain prognoses of the future to a discussion of human history. It unleashed a lively debate on the origins of global warming[13] and on the question of whether the long-term rise in greenhouse gases since the emergence of agriculture already prevented a new ice age in the more distant past.[14] In the course of the debate it became clear that the issues were so fundamental that they could not be so easily set aside.[15]

Nature conservation or human conservation?

Historiography and climatology received a fresh impetus from the moon landing of 1969, which made visible the great fragility of the Earth system. Natural and environmental conservation gained new significance and even acquired the character of an industry. Environmental associations and institutions launched campaigns scarcely less professional than those of transnational corporations. Since the 1990s, moreover, fear of global warming has pushed into the background earlier objects of eco-anxiety such as the *ozone hole* or the *death of the forests*. For the first time, it is not only industry but any final consumer that stands in the stocks. Virtually every inhabitant of the planet is guilty: the South African bushman who uses fire to clear land or to hunt; the Argentinian *ranchero* whose cattle produce methane; the rice-grower in Bali, and the Chinese banker who does business in an air-conditioned office.

When environmental or climatic conservation is spoken of in this context, we need to keep in mind what is really at issue. The earth has existed for more than five billion years, and there is a lot to suggest that it will last as long again, whatever human beings get up to on it. Throughout this time climate changes have been the rule, ranging from the hellishly hot planet ('the Hadean age') to 'snowball Earth'. For most of these billions of years it was hotter on earth than it is today. Only in the last few millions of years did the climate become more variable, either considerably warmer or – as a rule – much colder than today. Every change in climate has consequences for life on earth. But nature is not a moral system. Some species of plants and animals fare better under warm conditions, others under cold; some need more moisture, others less. As far as nature is concerned, ecosystem changes are neutral: that which harms one species offers advantages to another. Who would appoint himself as judge in these matters?

Natural conservation efforts are inherently conservative: what environmentalists seek to preserve is not 'nature' but a familiar kind of nature, an ecological state that is as much or as little 'natural' as any other. 'Natural conservation' is less about nature than about human well-being. And a good dose of inconsistency is apparent in the fact that many ecologically aware Central Europeans who fear global warming like to spend their holiday in warm parts of the world to escape the rain and the cold back home. The return of nature – for example, in the shape of 'Bruno', the brown bear, in the summer of 2006 – mainly triggers defensive reactions among those directly affected. Things are no different with regard to the climate. The big words 'climate protection' merely cover up the fear of change. A huge species diversity will spread in regions that have so far been at a disadvantage: high mountains and polar regions. Overspecialized species will become extinct. This is a question of evolution, not of morality.

Note that it is not the need for natural conservation that is being disputed here. But we must be clear about what is to be preserved, and why. That it is a major priority to save species threatened with extinction probably makes sense to everyone. But we may ask whether polar bears are an endangered species because of global warming, or because the Arctic is being opened up through human settlement, agriculture and industry. From a safe distance, one may regret that bears and humans cannot live side by side. But a polar bear around your refuse bin is as dangerous as Bruno in your front garden. Arctic wildlife will be as endangered as animals are in Africa or the Amazon basin, and serious thought must be given to how their existence outside zoos can be ensured in coexistence with rapidly expanding human settlements.

The struggle against atmospheric pollution makes sense, quite regardless of whether waste gases have a greenhouse effect or whether aerosols contribute to cooling. But, on our densely populated planet, resettlement will no longer be as simple as it was in the Neolithic and will probably lead to conflicts. The world community must therefore focus on keeping climate change within limits. It must prepare for climate change (*adaptation*) and prevent excessive climate change (*mitigation*). It makes no sense to play off the two strategies against each other, as people sometimes do.[16]

Climate policy as the challenge of the twenty-first century

After a generation of research, there is broad consensus among scientists about the fact of global warming and its anthropogenic component.[17] Their differences concern the interpretation of the research findings. The 'optimistic ecology' of James Lovelock – his Gaia hypothesis compared the earth to a goddess who automatically regulates the natural thermostat[18] – no longer has any appeal. Many climatologists still hold the view that we are heading for an ice age. Richard B. Alley still asks whether the Little Ice Age, the coldest period in the last ten thousand years, was not the first step towards another great ice age, since none of the last few interglacials lasted longer than ten thousand years.[19] From this, the politically incorrect conclusion is already drawn that anthropogenic warming may be a blessing in disguise, since it counteracts the slide into an ice age. Others, however, ask whether the warming is not actually much more dramatic than we think, if it is true that it not only raises long-term average temperatures but also offsets insidious cooling.

Most researchers see global warming as the number one problem for coming generations. For practical day-to-day politics, it is not necessary to clarify how great the anthropological component is; the first need is to take action against the causes and the consequences of warming. Many possible measures are not too expensive and would make sense even in the absence of climate change: for example, the removal of hidden subsidies for the burning of fossil fuels, the reduction of unhealthy waste gases, forest conservation or better thermal insulation. Some of these measures do not have to be left up to governments or international organizations but could be implemented at regional or local level by individual firms and households. Many American cities have recognized this and pursue a climate policy of their own that differs from the Washington guidelines. On the other hand, cheap and cheerful changes are limited in their effectiveness, and there is an urgent need for international coordination of measures that will have a greater impact on the climate. Talks on these issues will prepare the international community for a time when even more decisive measures might prove necessary.[20]

As the example of the Little Ice Age shows, even relatively slight warming can cause a major change to the conditions of life. It will remain urgently necessary to curb the increase in carbon dioxide emissions by a sufficient amount at least to stabilize them worldwide at a high level. The German government's scientific advisory board on global changes in the environment has estimated this level at 450 ppm, which would mean a global increase of 2 degrees Celsius in mean ground temperature. Beneath that mark, it is thought, there would not be any great disasters. Above it there is scarcely any limit to what is imaginable.[21] So, politicians are facing a challenge. The fact that the problems cannot be solved at national level does not make things easier. But, if human beings already influence the climate of the earth system, they must find common answers to this challenge. 'Challenge' and 'response': these were the categories with the help of which Arnold Toynbee theorized the rise and fall of civilizations in his *Study of History*.[22] They still have meaning today. Climate change is the challenge of our generation. On our answer depends not the well-being of the planet but our own well-being.

The earth is dying? Not yet.

In a book that began with the formation of the solar system, the final conclusion more or less suggests itself. Human history is short by geological standards – what are 30,000 years measured against five billion? And what are the three generations that we plan ahead in comparison with the five billion years that geologists still give our planet? The Potsdam Institute for Climate Impact Research has proposed the following timetable of decline. In roughly 800 million years from now, the average temperature will have risen to 30 degrees Celsius, though with CO_2 levels well below today's and even below those of the last great ice age. In roughly 1.2 billion years the temperature will rise to 40 degrees, and in 1.6 billion years to 70 degrees. Then no photosynthesis will be possible, and life as we know it will lose its foundation. On the continents nothing will remain but bare rock. As soon as the average temperature rises above the boiling point of water, the oceans will vaporize. At even higher temperatures plate tectonics will cease. In 3.5 to 6 billion years from now, our central star will have inflated so much that temperatures on Earth will exceed 1000 degrees. In such conditions the atmosphere will escape and rocks will melt. Earth will end as it began: as a hellishly hot planet.[23]

But we are still a long way from there, and by comparison we prefer to speak of a slight warming today. The cultural history of the climate is full of examples in which cold and drought proved to be the great enemies of civilization. As we saw in the case of the Little Ice Age, small changes in average temperatures can have massive

214

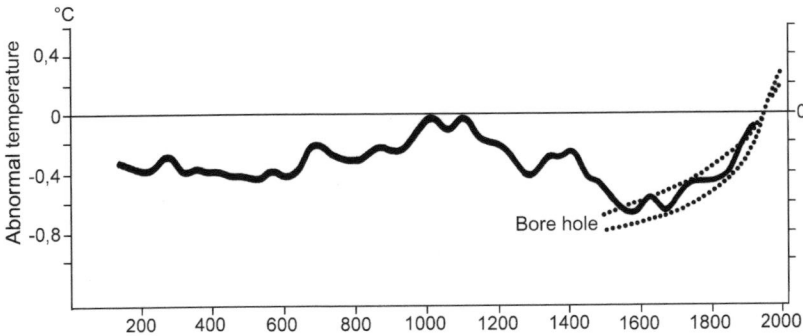

6.2 Evolution of global temperatures over past 2000 years, according to the latest ocean core drilling. What has happened to the hockey stick (see fig. 0.2)? And what does this mean for the good old theory that temperatures are determined by CO_2 content (see fig. 5.6)? This new calculation of temperatures is strikingly similar to the IPCC estimate of 1990 (see fig. 0.1).

consequences. Fortune and misfortune are then unevenly distributed. While Renaissance culture was flourishing in Italy, the Vikings were dying out in Greenland. During the 'summer of the century' in 2003, several thousand old people died in France from extreme heat and dehydration. Better care by the community could have prevented that from happening: there was nothing comparable in Austria, Germany or Switzerland, for example. Nor did the heat wave hold any terrors for people in air-conditioned rooms or on the beach. Global warming will require some adaptation and bring about some change. At the same time, we would like to hear the other side of the coin: how many fewer people die, lose their job or escape illness because the winters are milder. Harald Martenstein accurately sums up the environmentalist dilemma: 'People scream bloody murder when the number of species shrinks in one part of the world. But they also feel unhappy when the number of species grows in another part of the world.'[24]

Joachim Radkau writes of his colleagues: 'For many historians of the environment, the most discomfiting unknown is the climatic factor. For those who are seeking a moral and a message in environmental history, the climate is a disturbing factor devoid of meaning – at least for the longest period in history when human influence on the climate can be ruled out.'[25] But even for the period of anthropogenic warming it is not so easy to solve the moral question. Latter-day

215

priests speak of ecosins – but stockbreeders in Argentina or rice farmers in Indonesia belong in the same category of climate sinners as Texan oil corporations or operators of coal-fired power stations in China. Ideologues of the guilt culture demand not only remorse and atonement but also punishment in the name of the victims of climate change. Yet, from a legal standpoint, that could probably happen only if there was universal agreement or an eco-Stalinist world government – that is, hopefully never. The largest emitters are not yet even involved in emission trading. Sensible countries are already reducing their emissions, and the circle of them will grow wider.

We should not expect miracles, if climatologists are right about the inertia of the climate system. The earth will still continue to grow warmer even if every country behaves in model fashion and dramatically reduces its waste gases. This may well be a vexing thought. But it is not as bad as the older predictions of an imminent ice age: cooling has always resulted in major social upheavals, whereas warming has sometimes led to a blossoming of culture. If we can learn anything from the history of culture, it is that, even if humans were 'children of the Ice Age', civilization was a product of climatic warming. The Neolithic Revolution and the rise of ancient civilizations became possible in periods when it was somewhat warmer than it is today. If the IPCC's latest predictions are accurate, those levels will be reached again at some point in the twenty-first century. Then the Alpine glaciers will melt, but not those of the Antarctic. We will save on heating costs and use less fossil energy. What will become of the deserts? Will they really spread? During the Atlantic period, more water circulated in the atmosphere and the Sahara was fertile (see figure 2.2).

The future is hard to foresee. Serious scientists should refrain from slipping into the role of Nostradamus. Computer simulations cannot be better than the premises that guided the input of data: they show what is expected to happen, not the actual future. The history of the sciences is also a history of false theories and wrong predictions. It is interesting to become acquainted with the imprecision of scientific dating methods. 'Exact' datings with the C-14 or other physics-based methods must be 'calibrated' before they can become serviceable. Or, to put it more clearly: only historical records can put the 'exact' sciences on the right track. Academics in the humanities are not used to such a degree of imprecision. Although natural scientists usually estimate such dates with an expected variation of plus or minus one hundred years, historians can often pinpoint not only the exact day but sometimes the hour or even the very minute. We should be under no illusion about the exactitude of the natural sciences.

A 'cultural history of the climate' – and a concern with the cultural or social consequences of climate change – implies knowledge of the methodological premises of cultural studies.[26] It also means taking seriously data acquired not from ice and mud but from the archives of society. One is continually struck by the fruitfulness of a combination of historical and scientific methods.[27] Cultural history shows that the climate has always been in change and that society has always had to react to it. Apocalyptic prognoses have never been useful in such contexts, and we do not have to think back to the witch persecutions or the fall of ancient Egyptian dynasties to realize this. We need only take the measures planned in the 1970s to combat global cooling and compare them with the ones mooted today to combat global warming. Climatologists are advised to exercise moderation in speaking about climate history, and caution when it is a question of culture and society.

The climate is changing. The climate has always changed. How we react to it is a cultural question, and a knowledge of history can be of some help. Climate changes have often been experienced as threatening. False prophets and moral entrepreneurs have constantly tried to profit from this. We cannot leave the 'interpretation' of climate change to people ignorant of cultural history. Human beings are not animals that can only passively watch changes in their habitat. In recent history, climate change led to some positive developments. If the changes taking place today prove to be long-term – which is how things look at the moment – we can only advise everyone to keep calm. The world will not come to an end. If it becomes warmer, we will get used to it. Let us recall a classical Latin saying: *Tempora mutantur, et nos mutamur in illis*. Times change and we change in them.

NOTES

PREFACE

1   Work on this project proved more demanding than expected. Nowadays many disciplines make a contribution to our knowledge, and it takes time, patience and support to come to grips with their paradigms. Cooperation with Professors Hartmut Lehmann and Christian Pfister, and with participants in our climate conference at the Max Planck Institute for History, was a source of inspiration. I am especially grateful to my university, and to its presidents Professor Margaret Wintermantel and Professor Volker Linneweber, for their granting of a semester of research leave and continuing financial support for the post in cultural-historical studies. Of my colleagues, I would like to thank Mareike Kern, Annika Lauer and Stephan Rosenke, who have assisted this project with thoughts of their own. To my family I must apologize that I was so taken up with the topic, even during the recent vacation in the summer of the century. But at the time it was impossible to open a paper without seeing its importance confirmed.

INTRODUCTION

1   James Lovelock, *Gaia: A New Look at Life on Earth*, Oxford 1979.
2   Anon, 'Climate of Distrust', *Nature* 436 (2005), p. 1.
3   James J. McCarthy et al. (eds), *Climate Change 2001: Impact, Adaptation, and Vulnerability. Contribution of the Working*

Group II to the Third Assessment of the IPCC, Cambridge 2001.

4 Spencer Edward Abraham, 'The Bush Administration's Approach to Climate Change', *Science* 305 (2004), pp. 616–17.

5 Michael E. Mann/Raymond S. Bradley/Malcolm K. Hughes, 'Global-Scale Temperature Patterns and Climate Forcing over the Past Six Centuries', *Nature* 392 (1998), pp. 779–87.

6 Michael E. Mann et al., 'Northern Hemisphere Temperatures during the Past Millennium: Interferences, Uncertainties, and Limitations', *Geophysical Research Letters* 26 (1999), pp. 759–62.

7 *Brockhaus Enzyklopädie in 30 Bänden*, 21st edn, vol. 16, Mannheim 2006, p. 127.

8 John T. Houghton et al. (eds), *Climate Change 1995: The Science of Climate Change. Contribution of Working Group I to the Second Assessment of the IPCC*, Cambridge 1996.

9 Robert F. Kennedy, *Crimes against Nature*, New York 2004, p. 46.

10 Paul Andrew Mayewski and Frank White, *The Ice Chronicles*, London 2002, pp. 44–9.

11 Stephen H. Schneider, *Global Warming*, New York 1990. In the 1960s Schneider had defended the thesis of global cooling; his conversion to global warming came in the early 1970s.

12 Quoted in: *The Economist* editorial, 2 February 2002. Cf. Stephen Schneider, *Global Warming*, N4 1990 p. xi: '. . . capture the public's attention, if not its imagination, in order to create positive change. Often this means simple and dramatic statements – the easiest way to get media coverage.'

13 Stefan Rahmstorf, 'Klimawandel – rote Karte für die Leugner', in *Bild der Wissenschaft* (2003) – PDF download from the Rahmstorf home page: www.pik-potsdam.de.

14 Ulrich Cubasch and Dieter Kasang, *Anthropogener Klimawandel*, Gotha 2000.

15 Quirin Schiermeier, 'Past Climate Comes into Focus but Warm Forecast Stays Put', *Nature* 433 (2005), pp. 562f.

16 Christopher Schrader, 'Der Wandel bleibt. Ein politischer Großangriff auf die Klimaforschung endet als Scharmützel', *Süddeutsche Zeitung*, 4 August 2006, p. 16.

17 Hermann Flohn, *Das Problem der Klimaveränderungen in Vergangenheit und Zukunft*, Darmstadt 1985, p. 120.

1 WHAT DO WE KNOW ABOUT THE CLIMATE?

1 Rolf Meissner, *The Little Book of Planet Earth*, New York 2002, pp. 81ff.
2 John Imbrie and Katherine Palmer-Imbrie, *Ice Ages*, London 1979, pp. 135ff.
3 Cesare Emiliani, 'Pleistocene Temperatures', *Journal of Geology* 63 (1955), pp. 538–78.
4 In 1960 Libby received the Nobel Prize for his invention of the radiocarbon method.
5 Wilhelm Lauer and Jörg Bendix, *Klimatologie*, Brunswick 2004, pp. 280f.
6 Eugen Seibold, *Das Gedächtnis des Meeres: Boden, Wasser, Leben, Klima*, Munich 1991.
7 Richard Alley, *The Two-Mile Time Machine*, Princeton 2002.
8 Claus U. Hammer et al., 'Past Volcanism and Climate Revealed by Greenland Ice Cores', *Journal of Volcanology and Geothermal Research* 11 (1981), pp. 3–10.
9 Willi Dansgaard et al., 'One Thousand Centuries of Climatic Record from Camp Century on the Greenland Ice Sheet', *Science* 166 (1969), pp. 377–81.
10 Debra A. Meese et al., 'The Accumulation of Record from GISP2 Core as an Indicator of Climate Change throughout the Holocene', *Science* 266 (1994), pp. 1680–5.
11 J. R. Petit et al., 'Climate and Atmospheric History of the Past 420,000 Years from the Vostok Ice Core, Antarctica', *Nature* 399 (1999), pp. 429–36.
12 EPICA, 'Eight Glacial Cycles from an Antarctic Ice Core', *Nature* 429 (2004), pp. 623–8.
13 Paul Andrew Mayewski and Frank White, *The Ice Chronicles*, London 2002.
14 Stefan Winkler, *Von der 'Kleinen Eiszeit' zum 'globalen Gletscherrückzug'*, Stuttgart 2002.
15 Fritz H. Schweingruber, *Tree Rings*, Boston 1988.
16 George J. Gumerman, *The Anasazi in a Changing Environment*, New York 1988, p. 279.
17 Regiomontanus, *Ephemerides ab anno 1475–1506*, Nuremberg 1474.
18 Christian Pfister, *Wetternachhersage*, Berne 1999, pp. 18f.
19 Jean M. Grove and A. Battagel, 'Tax Records from Western Norway as an Index of the Little Ice Age Environmental and

Economic Deterioration', *Climatic Change* 3 (1983), pp. 265–82.

20 Walter Bauernfeind and Ulrich Woitek, 'The Influence of Climatic Change on Price Fluctuations in Germany during the 16th Century Price Revolution', *Climatic Change* 43 (1999), pp. 303–21.

21 Wilhelm Lauer and P. Frankenberg, *Zur Rekonstruktion des Klimas im Bereich der Rheinpfalz seit Mitte des 16. Jahrhunderts mit Hilfe der Weinquantität und Weinqualität*, Stuttgart 1986.

22 Christian Pfister, *Klimageschichte der Schweiz 1525–1860*, Berne/Stuttgart 1988.

23 Rüdiger Glaser, *Klimageschichte Mitteleuropas*, Darmstadt 2001.

24 Christian Pfister, *Wetternachhersage*, Berne 1999.

25 Karl Brunner, 'Ein buntes Klimaarchiv – Malerei, Graphik und Kartographie als Klimazeugen', *Naturwissenschaftliche Rundschau* 56 (2003), pp. 181–6.

26 Gordon Manley, 'Central England Temperatures: Monthly Means 1659 to 1973', *Quarterly Journal of the Royal Meteorological Society* 100 (1974), pp. 389–405.

27 Christian Pfister, *Wetternachhersage*, Berne 1999, pp. 26–9.

28 Transliteration from the Cyrillic script gives different versions of the name in published works: Milankovitch or Milankovic in the United States, Milankovitsch in Germany.

29 W. H. Berger, T. Bickert, E. Jansen, M. Yasuda and G. Werfer, 'Das Klima im Quartär – Rekonstruktion aus Tiefseesedimenten mit Hilfe der Milankovitch-Theorie', *Geowissenschaften* 12 (1994), pp. 258–66.

30 N. H. Shackleton and N. D. Opdyke, 'Oxygen Isotope and Palaeomagnetic Stratigraphy of Equatorial Pacific Core V28–238', *Journal of Quaternary Research* 3 (1991), pp. 39–55.

31 Paul Andrew Mayewski and Frank White, *The Ice Chronicles*, London 2002, pp. 97–110.

32 Jean-Marc Barnola et al., 'Reconstruction des variations du CO_2 atmosphérique en relation avec le climat au cours des derniers 160 000 ans à partir de la carotte de Vostok (Antarctique)', in Burkhard Frenzel (ed.), *Klimageschichtliche Probleme der letzten 130 000 Jahren*, Stuttgart/New York 1991, pp. 225–30.

33 Stefan Rahmstorf and Hans Joachim Schellhuber, *Der Klimawandel*, Munich 2006, pp. 17f.

34 Nico Stehr and Hans von Storch, *Klima, Wetter, Mensch*, Munich 1999, p. 73.
35 Steven M. Stanley, *Extinction*, New York 1987.
36 Hugh C. Owen, *Atlas of Continental Displacement, 200 Million Years to the Present*, Cambridge 1983.
37 Josef Klostermann, *Das Klima im Eiszeitalter*, Stuttgart 1999, pp. 224f.
38 Henry Stommel and Elizabeth Stommel, *Volcano Weather*, Newport 1983.
39 Tom Simkin and Lee Siebert, *Volcanoes of the World: A Regional Directory, Gazetteer, and Chronology of Volcanism during the Last 10,000 Years*, Tucson, AZ 1994, pp. 23–34.
40 Michael R. Rampino et al., 'Volcanic Winters', *Annual Review of Earth and Planetary Sciences* 16 (1988), pp. 73–99.
41 Rolf Meissner, *The Little Book of Planet Earth*, New York 2002, pp. 7, 38.
42 *Brockhaus Enzyklopädie in 30 Bänden*, 21st edn, vol. 10, Mannheim 2006, p. 498, using the appellations established in 2004 by the International Commission on Stratigraphy.
43 Peter Hupfer and Wilhelm Kuttler (eds), *Witterung und Klima*, Wiesbaden 2005, p. 280.
44 Rolf Meissner, *The Little Book of Planet Earth*, New York 2002, pp. 71ff.
45 Lazarus J. Salop (ed.), *Geological Evolution of the Earth during the Precambrian*, Berlin 1983.
46 Rolf Meissner, *The Little Book of Planet Earth*, New York 2002, p. 79.
47 Gabrielle Walker, *Snowball Earth*, New York 2003.
48 Consisting of the Cambrian, Ordovician, Silurian, Devonian, Carboniferous and Permian periods.
49 Consisting of the Triassic, Jurassic and Cretaceous periods.
50 Consisting of the Palaeocene, Eocene, Oligocene, Miocene, Pliocene, Pleistocene and Holocene epochs. The traditional 'Quaternary-Tertiary' periodization places the break after the Pliocene (1.8–2 million years ago). But the newer Palaeogene-Neogene periodization places the beginning of the Neogene period after the Oligocene (24 million years ago).
51 Rolf Meissner, *The Little Book of Planet Earth*, New York 2002, p. 79.
52 P. M. Sheehan, 'The Late Ordovician Mass Extinction', *Annual Review of Earth and Planetary Sciences* 29 (2001), pp. 331–64.

53 M. M. Joachimski and W. Buggisch, 'Condont Apatite Delta O18 Signatures Indicate Climatic Cooling as a Trigger of the Late Devonian Mass Extinction', *Geology* 30 (2002), pp. 711–14.

54 Douglas H. Erwin, *The Great Paleozoic Crisis: Life and Death in the Permian*, New York 1993.

55 G. Bloos, 'Untergang und Überleben am Ende der Trias-Zeit', in Wolfgang Hansch (ed.), *Katastrophen der Erdgeschichte – Wendezeiten des Lebens*, Heilbronn 2003, pp. 128–43.

56 J. David Archibald, *Dinosaur Extinction and the End of an Era*, New York 1996.

57 Steven M. Stanley, *Extinction*, New York 1987, pp. 91–107: quotation, p. 91.

58 W. M. Kürschner and H. Visscher, 'Das Massenaussterben an der Perm/Trias Grenze', in Wolfgang Hansch (ed.), *Katastrophen der Erdgeschichte – Wendezeiten des Lebens*, Heilbronn 2003, pp. 118–27.

59 Stanley, *Extinction*, pp. 109–31.

60 V. Moosbrugger, 'Das große Sterben vor 65 Millionen Jahren', in Wolfgang Hansch (ed.), *Katastrophen der Erdgeschichte – Wendezeiten des Lebens*, Heilbronn 2003, pp. 144–53.

61 Kenneth J. Hsü, *The Great Dying*, New York 1986.

62 Stanley, *Extinction*, pp. 47, 133–72.

63 According to the traditional nomenclature, the Cenozoic 'era' breaks down into two periods: the Tertiary (65 to 2 million years) and the Quaternary (from 2 million years to the present, consisting of the Pleistocene and Holocene). According to the new periodization, the same period of time is divided into the Palaeogene (65–23 million years) and the Neogene (23 million years to the present). The Palaeogene 'epochs' are the Palaeocene, Eocene and Oligocene, and the Neogene epochs the Miocene, Pliocene, Pleistocene and Holocene.

64 Rolf Meissner, *The Little Book of Planet Earth*, New York 2002, p. 166.

65 Hans-Jürgen Müller-Beck, *Die Eiszeiten*, Munich 2005, p. 36.

66 *Brockhaus Enzyklopädie*, vol. 7, pp. 646–9.

67 Müller-Beck, *Die Eiszeiten*, pp. 60–8.

68 Josef Klostermann, *Das Klima im Eiszeitalter*, Stuttgart 1999, pp. 192f.

69 Richard Leakey and Roger Lewin, *Origins*, London 1977, pp. 39f., 96.

70 Josef H. Reichholf, *Das Rätsel der Menschenwerdung*, Munich 2004, pp. 142–9.
71 Richard Leakey and Roger Lewin, *Origins*, London 1977.
72 Hans-Jürgen Müller-Beck, *Die Eiszeiten*, Munich 2005, p. 70.
73 Goran Burenhult, 'Towards Homo Sapiens: Habilines, Erectines, and Neanderthals', in Goran Burenhult (ed.), *The First Humans: Human Origins and History to 10,000 BC*, New York 1993, pp. 55–73.
74 R. Said and Hugues Faure, 'Chronological Framework: African Pluvial and Glacial Epochs', in Joseph Ki-Zerbo (ed.), *General History of Africa, I: Methodology and African Prehistory*, Paris 1989, pp. 146–66.

2 GLOBAL WARMING: THE HOLOCENE

1 John Gribbin and Mary Gribbin, *Children of the Ice: Climate and Human Origins*, Oxford 1990.
2 William I. Rose and Craig A. Chesner, 'Worldwide Dispersal of Ash and Gases from Earth's Largest Known Eruption: Toba, Sumatra, 75ka', *Palaeogeography, Palaeoclimatology, Palaeoecology* 89 (1990), pp. 269–75.
3 Michael R. Rampino and Stanley H. Ambrose, 'Volcanic Winter in the Garden of Eden: The Toba Supereruption and the Late Pleistocene Human Population Crash', in Floyd W. McCoy and Grant Heiken (eds), *Volcanic Hazards and Disasters in Human Antiquity*, Boulder, CO 2000, pp. 71–82.
4 Ann Gibbons, 'Pleistocene Population Explosions', *Science* 262 (1993), pp. 27–8.
5 Stanley H. Ambrose, 'Late Pleistocene Human Population Bottlenecks, Volcanic Winter, and Differentiation of Modern Humans', *Journal of Human Evolution* 34 (1982), pp. 623–51.
6 Michael R. Rampino and Stephen Self, 'Volcanic Winter and Accelerated Glaciation following the Toba Supereruption', *Nature* 359 (1992), pp. 50–2.
7 R. Said and Hugues Faure, 'Chronological Framework: African Pluvial and Glacial Epochs', in Joseph Ki-Zerbo (ed.), *General History of Africa, I: Methodology and African Prehistory*, Paris 1989, pp. 146–66.
8 Goran Burenhult, 'Towards Homo Sapiens: Habilines, Erectines, and Neanderthals', in Burenhult (ed.), *The First Humans:*

Human Origins and History to 10,000 BC, New York 1993, pp. 55–73.

9 Brian Fagan, *The Great Journey: the Peopling of Ancient America*, London 1987.

10 Goran Burenhult, 'Modern People in Africa and Europe', in Burenhult (ed.), *The First Humans: Human Origins and History to 10,000 BC*, New York 1993, pp. 77–95.

11 Gerhard Bosinski, 'Die Anfänge der Kunst: Das Jungpaläolithikum in Deutschland', in Manfred Nawroth (ed.), *Menschen, Zeiten, Räume: Archäologie in Deutschland*, Stuttgart 2003, pp. 113–18.

12 H. Kirchner, 'Ein archäologischer Beitrag zur Urgeschichte des Schamanismus', *Anthropos* 47 (1952), pp. 244–86.

13 Jean-Marie Chauvet, *Grotte Chauvet*, Sigmaringen 1995.

14 Goran Burenhult, 'Modern People in Africa and Europe', in Burenhult (ed.), *The First Humans: Human Origins and History to 10,000 BC*, New York 1993, pp. 77–95.

15 Ibid.

16 Paul S. Martin, 'Prehistoric Overkill: The Global Model', in Paul S. Martin and R. G. Klein (eds), *Quaternary Extinctions: A Prehistorical Revolution*, Tucson, AZ 1984, pp. 354–403.

17 R. Musil, 'Das Aussterben der Pleistozänen Großsäuger', in W. Hansch (ed.), *Katastrophen in der Weltgeschichte. Wendezeiten des Lebens. Museo 19* (2003), pp. 154–65.

18 Max Frisch, *Man in the Holocene: a Story*, London 1980.

19 Stephen H. Schneider and Randi Londer, *The Co-evolution of Climate and Life*, San Francisco 1984, p. 92.

20 Thomas J. Cronin, *Principles of Palaeoclimatology*, New York 1999, p. 259.

21 Georg W. Oesterdiekhoff, 'Geographische Bedingungen der Weltgeschichte. Die Wechselwirkung von Klima, Bevölkerungswachstum und Landnutzung in der Evolution der Hochkulturen', *Zeitschrift für Agrargeschichte und Agrarsoziologie* 47 (1999), pp. 123–32 (here p. 125).

22 Gerhard Bosinski, 'Die Anfänge der Kunst: Das Jungpaläolithikum in Deutschland', in Manfred Nawroth (ed.), *Menschen, Zeiten, Räume: Archäologie in Deutschland*, Stuttgart 2003, pp. 113–18.

23 Klaus Schmidt, *Sie bauten die ersten Tempel*, Munich 2006.

24 Stephen Mithen, *After the Ice: A Global Human History, 20,000–500 BC*, London 2003, pp. 56–61.

25 Achim Brauer et al., 'High Resolution Sediment and Vegetation Responses to Younger Dryas Climate Change in Varved Lake Sediments from Meerfelder Maar, Germany', *Quaternary Science Review* 18 (1999), pp. 321–9.

26 Michael Baales, 'Zwischen Kalt und Warm: Das Spätpaläolithikum in Deutschland', in Manfred Nawroth (ed.), *Menschen, Zeiten, Räume: Archäologie in Deutschland*, Stuttgart 2003, pp. 121–3.

27 Daniel Schwemer, *Die Wettergottgestalten Mesopotamiens und Nordsyriens im Zeitalter der Keilschriftkulturen*, Wiesbaden 2001, pp. 11–16.

28 John F. B. Mitchell, 'Greenhouse Warming: Is the Mid-Holocene a Good Analogue?', *Journal of Climate* 3 (1990), pp. 1177–92.

29 Stephen Mithen, *After the Ice: A Global Human History, 20,000–500 BC*, pp. 153f.

30 Claus Joachim Kind, 'Die letzten Jäger und Sammler: Die Mittelsteinzeit', in Nawroth (ed.), *Menschen, Zeiten, Räume: Archäologie in Deutschland*, pp. 124–7. The datings appear here with a difference of two thousand years, probably because of a confusion between BP (before present) and AD dates. Cf. Wilhelm Lauer and Jörg Bendix, *Klimatologie*, Brunswick 2004, p. 285.

31 Alfred Rust, 'Der primitive Mensch', in Alfred Heuss and Golo Mann, *Propyläen Weltgeschichte: Eine Universalgeschichte*, 10 vols., West Berlin 1960–4, vol. 1, pp. 135–226, 218.

32 Hansjörg Küster, *Geschichte der Landschaft in Mitteleuropa*, Munich 1995, pp. 66ff.

33 Helmut Jäger, however (*Einführung in die Umweltgeschichte*, Darmstadt 1994, p. 25), dates the Atlantic to the period from 5500 to 250 BC.

34 R. Said and Hugues Faure, 'Chronological Framework: African Pluvial and Glacial Epochs', in Joseph Ki-Zerbo (ed.), *General History of Africa, I: Methodology and African Prehistory*, Paris 1989, pp. 156ff.

35 Wilhelm Lauer and Jörg Bendix, *Klimatologie*, Brunswick 2004, pp. 286f.

36 Andreas Zimmermann, 'Der Beginn der Landwirtschaft in Mitteleuropa', in Nawroth (ed.), *Menschen, Zeiten, Räume: Archäologie in Deutschland*, pp. 133–4.

37 Stephen Mithen, *After the Ice: A Global Human History, 20,000–500 BC*, London 2003, pp. 62–79.

38 Norbert Benecke, *Der Mensch und seine Haustiere*, Stuttgart 1994, pp. 77–94, 143ff.

39 Neil Roberts, *The Holocene: An Environmental History*, Oxford 1998.

40 Gina L. Barnes and Peter Bellwood, 'Stone Age Farmers in Southern and Eastern Asia: China, Korea and Japan', in *The Illustrated History of Humankind*, Sydney 1993, pp. 134–7, 140–2.

41 Hubert H. Lamb, *Climate, History and the Modern World*, 2nd edn, London 1995, pp. 124ff.

42 Herbert Jankuhn, 'Der Ursprung der Hochkulturen', in Heuss and Mann (eds), *Propyläen Weltgeschichte*, vol. 2, pp. 573–600, esp. pp. 576f.

43 Georg W. Oesterdiekhoff, 'Geographische Bedingungen der Weltgeschichte. Die Wechselwirkungen von Klima, Bevölkerungswachstum und Landnutzung in der Evolution der Hochkulturen', *Zeitschrift für Agrargeschichte und Agrarsoziologie* 47 (1999), pp. 123–32; here, p. 126.

44 Benecke, *Der Mensch und seine Haustiere*, Stuttgart 1994.

45 Martin A. J. Williams and Hugues Faure (eds), *The Sahara and the Nile*, Rotterdam 1980.

46 Richard G. Klein, 'Hunter-gatherers and Farmers in Africa', in Goran Burenhult (ed.), *People of the Stone Age*, San Francisco 1993, pp. 39–57, esp. p. 43.

47 'Ägypten', in *Brockhaus Enzyklopädie in 30 Bänden*, 21st edn, vol. 1, Mannheim 2006, pp. 335–58, esp. pp. 341ff.

48 'Nil', in *Der Neue Pauly: Enzyklopädie der Antike*, vol. 8 (2000), esp. pp. 942–4.

49 Irenäus Matuschik, Johannes Müller and Helmut Schlichtherle, 'Technik, Innovation und Wirtschaftswandel: Die späte Jungsteinzeit', in Nawroth (ed.), *Menschen, Zeiten, Räume: Archäologie in Deutschland*, pp. 156–61.

50 Konrad Spindler, *The Man in the Ice*, London 1994.

51 John Baines and Jaromir Málek, *Atlas of Ancient Egypt*, Oxford 1980, p. 14.

52 Barbara Bell, 'The Dark Ages in Ancient History: I: The First Dark Age in Egypt', *American Journal of Archeology* 75 (1971), pp. 1–26.

53 *Der Garten in Eden: 7 Jahrtausende Kunst und Kultur an Euphrat und Tigris*, exhibition catalogue, Mainz 1978, p. 8.

54 Claus Wilcke (ed.), *Das Lugalbandaepos*, Wiesbaden 1969, p. 119 – *The Electronic Text Corpus of Sumerian Literature*, http://etcsl.orinst.ox.ac.uk/cgi-bin/etcsl.cgi?text=t.1.8.2.2#.

55 Norman Yoffee, 'The Collapse of Ancient Mesopotamian States and Civilization', in Norman Yoffee and George L. Cowgill (eds), *The Collapse of Ancient States and Civilizations*, Tucson, AZ 1988, pp. 44–68.

56 Peter B. DeMenocal, 'Cultural Responses to Climate Change during the Late Holocene', *Science* 292 (2001), pp. 667–73, esp. pp. 669f.

57 H. M. Cullen et al., 'Climate Change and the Collapse of the Akkadian Empire: Evidence from Deep Sea', *Geology* 28 (2000), pp. 379–82.

58 Paul Andrew Mayewski and Frank White, *The Ice Chronicles*, London 2002, pp. 102f.

59 Herbert Kaufman, 'The Collapse of Ancient States and Civilizations as an Organizational Problem', in Yoffee and Cowgill (eds), *The Collapse of Ancient States and Civilizations*, pp. 219–35.

60 Harvey Weiss et al., 'The Genesis and Collapse of Third Millennium North Mesopotamian Civilization', *Science* 261 (1993), pp. 995–1004.

61 Daniel Schwemer, *Die Wettergottgestalten Mesopotamiens und Nordsyriens im Zeitalter der Keilschriftkulturen*, Wiesbaden 2001, p. 436.

62 Michael Jansen and G. Urban (eds), *Mohenjo Daro*, Leiden 1985.

63 G. Singh, 'The Indus Valley Culture', *Archeology and Physical Anthropology in Oceania* 6 (1971), pp. 177–89.

64 Hermann Kulke and Dieter Rothermund, *Geschichte Indiens*, Munich 1998, pp. 9–13, 25–44.

65 Barbara Bell, 'Climate and the History of Egypt: The Middle Kingdom', *American Journal of Archeology* 79 (1975), pp. 223–69.

66 Anthony F. Harding, *European Societies in the Bronze Age*, Cambridge 2000, pp. 197–270.

67 Klaus-Dieter Jäger and Vojen Lozek, 'Environmental Conditions and Land Cultivation during the Urnfield Bronze Age in Central Europe', in Anthony F. Harding (ed.), *Climatic Change in Later Prehistory*, Edinburgh 1982, pp. 162–78.

68 'Mykene', in *Der Neue Pauly: Enzyklopädie der Antike*, vol. 8 (2000), esp. pp. 571–87.

69 William A. Ward and Martha Sharp Joukowsky, *The Crisis Years: The 12ᵗʰ Century BC from beyond the Danube to the Tigris*, Dubuque, IA 1989.

70 Aristotle, *Meteorology* I:15, trans. by E. W. Webster, *Internet Classics Archive*, at http://classics.mit.edu/Aristotle/meteorology.1.i.html.

71 Reid A. Bryson et al., 'Drought and the Decline of Mycenae', *Antiquity* 48 (1974), pp. 46–50.

72 Reid A. Bryson and Thomas J. Murray, *Climates of Hunger: Mankind and the World's Changing Weather*, Madison, WI 1977, pp. 3–17.

73 Horst Klengel, 'Hungerjahre in Hatti', *Altorientalische Forschungen* I (1974), pp. 165–74.

74 'Hattusa', in *Der Neue Pauly: Enzyklopädie der Antike*, vol. 5 (1998), pp. 185–98.

75 'Juda und Israel', 'Judentum', in ibid., esp. pp. 1187–1200.

76 'Hadad und Jahwe', in ibid., esp. pp. 50–1, 841–3.

77 S. Grätz, *Der strafende Wettergott: Erwägungen zur Traditionsgeschichte des Adad-Fluchs im Alten Orient und im Alten Testament (= Bonner Biblische Beiträge)*, Bodenheim 1998.

78 See the keywords *Baal*, *Blitz* (lightning), *Dürre* (drought), *Fieber* (fever), *Finsternis* (darkness), *Flut* (flood), *Hagel* (hail), *Hunger* (famine), *Pest* (plague), *verdorren* (to wither), *vernichten* (to slay), *Wetter* (weather) and *Zorn* (wrath), in *Neue Konkordanz zur Einheitsübersetzung der Bibel*, edited by Franz Josef Schiersee, revised by Winfried Bader, Darmstadt 1996.

79 Hubert H. Lamb, 'Reconstruction of the Course of the Postglacial Climate over the World', in Anthony F. Harding (ed.), *Climatic Change in Later Prehistory*, Edinburgh 1982, pp. 11–32, diag. p. 30.

80 Kevin D. Pang et al., 'Climatic and Hydrologic Extremes in Early Chinese History: Possible Causes and Dates' (abstract), *Eos* 70 (1989), p. 1095.

81 William W. Hallo, 'From Bronze Age to Iron Age in Western Asia: Defining the Problem', in Ward and Joukowsky, *The Crisis Years: The 12ᵗʰ Century BC from beyond the Danube to the Tigris*, pp. 1–9.

82 B. Weiss, 'The Decline of Late Bronze Age Civilizations as a Possible Response to Climatic Change', *Climatic Change* 4 (1982), pp. 172–98.

83 Bernd Zolitschka et al., 'Humans and Climatic Impact on the Environment as Derived from Colluvial, Fluvial and Lacustrine

Archives – Examples from the Bronze Age to the Migration Period, Germany', *Quaternary Science Review* 22 (2003), pp. 81–100, esp. p. 97.

84 Günter Smolla, 'Der "Klimasturz" um 800 vor Chr. und seine Bedeutung für die Kulturentwicklung in Südwestdeutschland', in *Festschrift Peter Goessler*, Stuttgart 1954, pp. 168–86.

85 Bernd Zolitschka et al., 'Humans and Climatic Impact on the Environment . . .', p. 96.

86 Ludwig Pauli, *Die Alpen in Frühzeit und Mittelalter*, Munich 1981, pp. 42–5.

87 M. Stuiver and B. Becker, 'High Precision Decadal Calibration of Radiocarbon Time-Scale, AD 1950 – 6000 BC', *Radiocarbon* 35 (1993), pp. 35–66.

88 Baines and Málek, *Atlas of Ancient Egypt*, Oxford 1980, p. 49.

89 Wilhelm Lauer and Jörg Bendix, *Klimatologie*, Brunswick 2004, p. 287.

90 B. D. Shaw, 'Climate, Environment and History: The Case of Roman North Africa', in T. M. L. Wigley, M. J. Ingram and G. Farmer (eds), *Climate and History*, Cambridge 1981, pp. 379–403.

91 Hubert H. Lamb, *Climate, History and the Modern World*, London 1982, pp. 149, 159f.

92 Georg W. Oesterdiekhoff, 'Geographische Bedingungen der Weltgeschichte. Die Wechselwirkung von Klima, Bevölkerungswachstum und Landnutzung in der Evolution der Hochkulturen', *Zeitschrift für Agrargeschichte und Agrarsoziologie* 47 (1999), pp.123–32, p. 126.

93 Bernd Zolitschka et al., 'Humans and Climatic Impact on the Environment as Derived from Colluvial, Fluvial and Lacustrine Archives – Examples from the Bronze Age to the Migration Period, Germany', *Quaternary Science Review* 22 (2003), p. 98.

94 Christian-Dietrich Schönwiese, *Klimaänderungen*, Berlin 1995, p. 91.

95 Hubert H. Lamb, *Climate, History and the Modern World*, pp. 149ff.

96 Herbert Franke and Rolf Trauzettel, *Das Chinesische Kaiserreich*, Frankfurt/Main 1968, pp. 74–116.

97 Franz Altheim, *Die Geschichte der Hunnen*, 5 vols., Berlin 1962–1975.

98 Ulrich Fellmeth, *Brot und Politik*, Stuttgart 2001, pp. 156f.

99   Andreas Alföldi, *Studien zur Geschichte der Weltkrise des 3. Jahrhunderts*, Darmstadt 1967.
100 'Severinus von Noricum', in *Lexikon des Mittelalters* 7 (1999), esp. pp. 1805–6.
101 Eugippius, *The Life of Saint Severinus*, Cambridge, MA 1914, p. 23.
102 Hubert H. Lamb, *Climate, History and the Modern World*, p. 160.
103 Kenneth J. Hsü, *Klima macht Geschichte*, Zurich 2000, pp. 43f.
104 Heinrich Franke and Rolf Trauzettel, *Das Chinesische Kaiserreich*, Frankfurt/Main 1968, pp. 117–87.
105 Helmut Jäger, *Einführung in die Umweltgeschichte*, Darmstadt 1994, p. 26.
106 Christian-Dietrich Schönwiese, *Klimaänderungen*, Berlin 1995, pp. 81–6 *passim*.
107 Hubert H. Lamb, 'Reconstruction of the Course of the Postglacial Climate over the World', in Anthony F. Harding (ed.), *Climatic Change in Later Prehistory*, Edinburgh 1982, pp. 11–32, summary on p. 21.
108 Hubert H. Lamb, *Climate, History and the Modern World*, pp. 158f.
109 Christian-Dietrich Schönwiese, *Klimaänderungen*, Berlin 1995, pp. 83, 86.
110 Franz Georg Maier, *Die Verwandlung der Mittelmeerwelt*, Frankfurt/Main 1968.
111 Zolitschka et al., 'Humans and Climatic Impact on the Environment as Derived from Colluvial, Fluvial and Lacustrine Archives . . .', pp. 96ff.
112 Richard B. Stothers, 'Mystery Cloud of AD 536', *Nature* 307 (1984), pp. 344–5.
113 Gregory of Tours, *The History of the Franks*, Harmondsworth 1974.
114 Georges Duby, *The Early Growth of the European Economy*, London 1974, pp. 11f.
115 Ibid., pp. 6, 250–1.
116 Pierre Riché, *Daily Life in the World of Charlemagne*, rev. edn, Philadelphia, PA 1988, p. 25.
117 Ibid., pp. 250–1.
118 Arno Borst, *Lebensformen im Mittelalter*, Frankfurt/Main 1973, pp. 373f.
119 Ibid., p. 374.

120 *Annales Regni Francorum*, ed. by Reinhold Rau, pp. 124f. – quoted from Hans-Werner Goetz, *Life in the Middle Ages*, Notre Dame, IN 1993, pp. 15–6; translation slightly modified.

121 Goetz, *Life in the Middle Ages*, p. 17.

122 Hubert H. Lamb, *Climate, History and the Modern World*, p. 163.

123 D. A. Hodell et al., 'Possible Role of Climate in the Collapse of Classic Maya Civilization', *Nature* 375 (1995), pp. 391–4.

124 T. Patrick Culbert, 'The Collapse of the Classic Maya Civilization', in Norman Yoffee and George L. Cowgill (eds), *The Collapse of Ancient States and Civilizations*, Tucson, AZ 1988, pp. 69–101.

125 Richardson B. Gill, *The Great Maya Droughts: Water, Life and Death*, Albuquerque, NM 2000.

126 Larry C. Peterson and Gerald H. Haug, 'Climate and the Collapse of Maya Civilization', *American Scientist* 93:44 (2005), pp. 322–9.

127 Gerald H. Haug et al., 'Southward Migration of the Intertropical Convergence Zone through the Holocene', *Science* 293 (2001), pp. 1304–8.

128 B. G. Hunt and T. I. Elliott, 'A Simulation of the Climatic Conditions Associated with the Collapse of the Maya Civilization', *Climatic Change* 69 (2005), pp. 393–407.

129 Henry F. Diaz and Vera Markgraf (eds), *El Niño: Historical and Paleoclimatic Aspects of the Southern Oscillation*, Cambridge 1992, p. 315.

130 Allison C. Paulsen, 'Environment and Empire: Climatic Factors in Pre-Historic Andean Culture Change', *World Archaeology* 8 (1976), pp. 121–32.

131 Yoshshito Shimada et al., 'Cultural Impacts of Severe Droughts in the pre-Historic Andes: Application of a 1500-Year Ice Core Precipitation Record', *World Archaeology* 22 (1991), pp. 247–70.

132 L. G. Thompson et al., 'Reconstructing Interannual Climate Variability from Tropical and Subtropical Ice-Core Records', in Diaz and Markgraf (eds), *El Niño*, pp. 295–322: here p. 318.

133 Roger Y. Anderson, 'Long-Term Changes in the Frequency of Occurrence of El Niño Events', in Diaz and Markgraf (eds), *El Niño*, pp. 193–200.

134 Hermann Flohn, *Das Problem der Klimaänderungen in Vergangenheit und Zukunft*, Darmstadt 1985, p. 131.

135 Hubert H. Lamb, 'The Early Medieval Warm Epoch and Its Sequel', *Palaeogeography, Palaeoclimatology, Palaeoecology* 1 (1965), pp. 13–37.

136 Malcolm K. Hughes and Henry F. Diaz, 'Was There a "Medieval Warm Period", and If So, Where and When?', *Climatic Change* 26 (1994), pp. 109–42.

137 Raymond S. Bradley et al., 'Climate in Medieval Time', *Science* 302 (2003), pp. 404–5.

138 Jean M. Grove and Roy Switsur, 'Glacial Geological Evidence for the Medieval Warm Period', *Climatic Change* 26 (1994), pp. 143–69.

139 Pierre Alexandre, *Le Climat en Europe au Moyen Âge*, Paris 1987, pp. 775–808.

140 J. L. Jirikowic and P. E. Damon, 'The Medieval Solar Activity Maximum', *Climatic Change* 26 (1994), pp. 309–16.

141 Hubert H. Lamb, *Climate, History and the Modern World*, p. 157.

142 Dario Camuffo, 'Freezing of the Venetian Lagoon Since the 9th Century AD in Comparison to the Climate of Western Europe and England', *Climatic Change* 10 (1987), pp. 43–66, esp. pp. 58–64.

143 Rüdiger Glaser, *Klimageschichte Mitteleuropas*, Darmstadt 2001, pp. 61–92.

144 Hubert H. Lamb, *Climate, History and the Modern World*, pp. 125ff.

145 Wilfried Weber, *Die Entwicklung der nördlichen Weinbaugrenze in Europa*, Trier 1980.

146 Hubert H. Lamb, *Climate, History and the Modern World*, pp. 170f.

147 Andreas Holmsen, *Norske Historie*, Oslo 1961.

148 Zhang De'er, 'Evidence for the Existence of the Medieval Warm Period in China', *Climatic Change* 26 (1994), pp. 289–97.

149 Hubert H. Lamb, *Climate, History and the Modern World*, p. 173.

150 Paul C. Buckland and Pat E. Wagner, 'Is There an Insect Signal for the "Little Ice Age"?', *Climatic Change* 48 (2001), pp. 137–49.

151 Hans-Werner Goetz, *Life in the Middle Ages*, Notre Dame, IN, p. 17.

152 Jan Dhondt, *Das frühe Mittelalter*, Frankfurt/Main 1968, pp. 272–9.

153 Jacques Le Goff, *Le Moyen Âge*, Paris 1962, pp. 113–21.

154 Christian-Dietrich Schönwiese, *Klimaänderungen*, Berlin 1995, p. 2.

155 Jacques Le Goff, *Time, Work, and Culture in the Middle Ages*, Chicago 1980.

156 C. T. Smith, *An Historical Geography of Western Europe*, 2nd edn, London 1969, pp. 129–82.

157 Jacques Le Goff, *Le Moyen Âge*, Paris 1962, pp. 130f.

158 Heinz Stoob, *Forschungen zum Städtewesen in Europa*, Cologne 1970.

159 James Graham-Campbell (ed.), *Cultural Atlas of the Viking World*, Oxford 1994, pp. 164–84.

160 Richard F. Tomasson, 'Millennium of Misery', *Population Studies* 31 (1977), pp. 405–27.

161 Willi Dansgaard et al., 'Climate Changes, Norsemen and Modern Man', *Nature* 255 (1975), pp. 24–8.

162 Graham-Campbell (ed.), *Cultural Atlas of the Viking World*, Oxford 1994.

3 GLOBAL COOLING: THE LITTLE ICE AGE

1 François E. Matthes, 'Report of Committee on Glaciers', *Transactions of the American Geophysical Union* 20 (1939), pp. 518–23.

2 François E. Matthes, 'The Little Ice Age of Historic Times', in F. Fryxel (ed.), *The Incomparable Valley: A Geological Interpretation of the Yosemite*, Berkeley 1950, pp. 151–60.

3 Gustaf Utterström, 'Climatic Fluctuations and Population Problems in Early Modern History', *Scandinavian Economic History Review* 3 (1955), pp. 3–47.

4 Eric J. Hobsbawm, 'The General Crisis of the European Economy in the 17th Century', *Past and Present* 5 (1954), pp. 33–53.

5 Trevor Aston (ed.), *Crisis in Europe 1560–1660*, London 1965.

6 Émile Durkheim, *The Rules of Sociological Method*, 8th edn, Chicago 1938.

7 Emmanuel Le Roy Ladurie, 'Histoire et climat', *Annales ESC* 14 (1959), pp. 3–34.

8 Peter Burke, *The French Historical Revolution: The Annales School, 1929–89*, Cambridge 1990.

9 Emmanuel Le Roy Ladurie, *Times of Feast, Times of Famine*, New York 1971.

10 Hubert H. Lamb, *Climate: Present, Past and Future*, 2 vols., London 1972/1977.

11 Christian Pfister, *Klimageschichte der Schweiz 1525–1860*, Berne 1988.

12 Rudolf Brádzil (ed.), *Climatic Change in the Historical and Instrumental Periods*, Brno 1990.

13 Rüdiger Glaser, 'Die Temperaturverhältnisse in Württemberg in der frühen Neuzeit', *Zeitschrift für Agrargeschichte und Agrarsoziologie* 38 (1990), pp. 129–44.

14 Christian Pfister, 'Klimawandel in der Geschichte Europas. Zur Entwicklung und zum Potential der Historischen Klimatologie', in Erich Lansteiner (ed.), *Klima Geschichten*, Vienna 2001, pp. 7–43.

15 Gisli Gunnarson, *A Study of Causal Relations in Climate and History*, Lund 1980, p. 7.

16 Hermann Flohn, *Das Problem der Klimaänderungen in Vergangenheit und Zukunft*, Darmstadt 1985, p. 126.

17 Pierre Alexandre, *Le Climat en Europe au Moyen Âge*, Paris 1987, pp. 807f.

18 Axel Steensberg, 'Archaeological Dating of Climatic Change in North Europe about AD 1300', *Nature* 168 (1951), pp. 672–4.

19 John A. Eddy, 'The "Maunder Minimum": Sunspots and Climate in the Reign of Louis XIV', in Geoffrey Parker and Lesley M. Smith (eds), *The General Crisis of the Seventeenth Century*, London 1978, pp. 226–8.

20 George C. Reid, 'Solar Forcing of Global Climate Change since the Mid-Seventeenth Century', *Climatic Change* 37 (1997), pp. 391–405.

21 Claus U. Hammer et al., 'Past Volcanism and Climate Revealed by Greenland Ice Cores', *Journal of Volcanology and Geothermal Research* 11 (1981), pp. 3–10.

22 Kevin D. Pang, 'Climatic Impact of the Mid-Fifteenth Century Kuwae Caldera Formation, as Reconstructed from Historical and Proxy Data', *Eos* 74 (1993), p. 106.

23 Chaochao Gao et al., 'The 1452 or 1453 AD Kuwae Eruption Signal Derived from Multiple Ice-Core Records: Greatest Volcanic Sulfate Event of the Past 700 Years', *Journal of Geophysical Research*, January 2006, pp. 1–29 (online version).

24 Anne S. Palmer et al., 'High Precision Dating of Volcanic Events (AD 1301–1995), Using Ice Core from Law Dome,

Antarctica', *Journal of Geophysical Research* 106 (2001), p. 1953.

25 Shanaka L. de Silva, J. Alzueta and G. Salas, 'The Socioeconomic Consequences of the AD 1600 Eruption of Huaynaputina, Southern Peru', in Floyd W. McCoy and Grant Heiken (eds), *Volcanic Hazards and Disasters in Human Antiquity*, Boulder, CO 2000, pp. 15–24.

26 Shanaka L. de Silva and Gregory A. Zielinski, 'Global Influence of the AD 1600 Eruption of Huaynaputina, Peru', *Nature* 393 (1998), pp. 455–8.

27 Keith R. Briffa et al., 'Influence of Volcanic Eruptions on Northern Hemisphere Summer Temperature over the Past 600 Years', *Nature* 393 (1998), pp. 450–5.

28 Jean M. Grove, *The Little Ice Age*, London/New York 1988.

29 Jean M. Grove, 'The Initiation of the Little Ice Age in Regions around the North Atlantic', *Climatic Change* 48 (2001), pp. 53–82.

30 Jean M. Grove, 'The Onset of the Little Ice Age', in Phil D. Jones et al., *History and Climate: Memories of the Future?*, Dordrecht 2001, pp. 153–87.

31 John F. Richards, *The Unending Frontier*, Berkeley 2003, pp. 79–82, maps on pp. 62–3.

32 Ignacio Olagüe, *La decadencia de España*, Madrid 1950, vol. 4, ch. 25.

33 Jean M. Grove and Annalisa Conterio, 'The Climate of Crete in the Sixteenth and Seventeenth Centuries', *Climatic Change* 30/2 (1995), pp. 223–47: here, p. 231.

34 'Von einem grosen tüfen Schnee und wie vil lüth erfroren und im schnee erstickt und umbkommen' [1571], in Matthias Senn (ed.), *Die Wickiana*, Zurich 1975, p. 187.

35 Hartmann Braun, *Nix altissima, d.i. Der große tieffe Schnee, so an etzlichen Orthen im Anfang des 1611 gefallen, in forma concionis für Augen gestellet, über Syrach 43, 14*, Darmstadt 1611.

36 Martin Pezold, *Gottes weisser Mann und Winterkleid/Das ist: Schöne Anmuthige Christliche Schneegedancken/von den überaußgrossen gefallen Schneen/sonderlich dieses 1624. Jahres: Mit vielen schönen geistlichen und weltlichen Historien versetzet/sampt einem Catalogo Wunder [. . .] Geschichten/so sich von Anno 400. bis auff das [. . .] 1614. Jahr begeben*, Jena 1624.

37 Emmanuel Le Roy Ladurie, 'Writing the History of Climate', in idem, *The Territory of the Historian*, Chicago 1979, pp. 287–91, here p. 287.

38 *Topographia Helvetiae*, Frankfurt 1654, pp. 31f.

39 Chia-cheng Chang, *The Reconstruction of Climate in China for Historical Times*, Beijing 1988.

40 Werner Dobras, *Wenn der ganze Bodensee zugefroren ist*, Konstanz 1983.

41 W. Gregory Monahan, *Years of Sorrows: The Great Famine of 1709 in Lyon*, Columbus, OH 1993, pp. 72f.

42 Fernand Braudel, *The Mediterranean and the Mediterranean World in the Age of Philip II*, vol. 1, London 1975, pp. 272ff.

43 Henry S. Lucas, 'The Great European Famine of 1315, 1316 and 1317', *Speculum* 5 (1930), pp. 343–77.

44 Dario Camuffo, 'Freezing of the Venetian Lagoon since the 9th Century AD in Comparison to the Climate of Western Europe and England', *Climatic Change* 10 (1987), pp. 43–66, here pp. 58–64.

45 Daniel Schaller, *Herold*, Magdeburg 1595, pp. 156f.

46 Erich Landsteiner, 'Wenig Brot und saurer Wein', in Wolfgang Behringer et al. (eds), *Kulturelle Konsequenzen der 'Kleinen Eiszeit'*, Göttingen 2005, pp. 87–148.

47 Wilfried Weber, *Die Entwicklung der nördlichen Weinbaugrenze in Europa*, Trier 1980.

48 Wilhelm Lauer and Peter Frankenberg, *Zur Rekonstruktion des Klimas im Bereich der Rheinpfalz seit Mitte des 16. Jahrhunderts mit Hilfe der Weinquantität und Weinqualität*, Stuttgart 1986.

49 *Das Buch Weinsberg: Kölner Denkwürdigkeiten aus dem 16. Jahrhundert*, vol. 5, Bonn 1926, p. 316.

50 Fernand Braudel, *The Mediterranean and the Mediterranean World in the Age of Philip II*, vol. 1, London 1975.

51 Christian Pfister, *Klimageschichte der Schweiz 1525–1860*, Berne 1988.

52 Harald Bugmann and Christian Pfister, 'Impacts of Interannual Climate Variability on Past and Future Forest Composition', *Regional Environmental Change* 1 (2000), pp. 112–25.

53 Schaller, *Herold*, pp. 156ff.

54 Brian Fagan, *The Little Ice Age*, New York 2000, pp. 69–77.

55 H. Haller, 'Die Thermikabhängigkeit des Bartgeiers *Gypaetus barbatus* als mögliche Mitursache für sein Aussterben in den Alpen', *Ornithologische Beobachtungen* 80 (1983), pp. 263–72.

56 'Korrespondenz Bullinger', in Matthias Senn (ed.), *Die Wickiana*, Zurich 1975, pp. 186f.

57 Bugmann and Pfister, 'Impacts of Interannual Climate Variability . . .', pp. 112–25.

58 Martin Körner, 'Geschichte und Soziologie interdisziplinär: Feld- und Schermäuse in Solothurn 1538–1543', in *Jahrbuch für Solothurnische Geschichte* 66 (1993), pp. 441–54.

59 James R. Busvine, *Insects, Hygiene and History*, London 1976; James C. Riley, 'Insects and the European Mortality Decline', *American Historical Review* 91 (1986), pp. 833–58.

60 Hans Zinsser, *Rats, Lice and History*, London 1935.

61 Johann Fischart, *Flöh Haz, Weiber Traz*, Strassburg 1573.

62 *Floia, cortum versicale, de flois schwartibus, illis deiriculis, quae omnes fere Minschos, Mannos, Weibras, Jungfras etc. behuppere et spitzibus suis schnaflis steckere et bitere solent. Autore Gripholdo Knickknackio ex Floilandia*, Year 1593.

63 Willi Dansgaard et al., 'Climate Changes, Norsemen and Modern Man', *Nature* 255 (1975), pp. 24–8.

64 Wolfgang Behringer, *Witches and Witch Hunts: A Global History*, Cambridge 2004.

65 James Graham-Campbell (ed.), *A Cultural Atlas of the Viking World*, Oxford 1994, pp. 222–3.

66 Jens Peder Hart Hansen et al., *The Greenland Mummies: The British Museum*, London 1991.

67 Bent Fredskild, 'Agriculture in a Marginal Area: South Greenland from the Norse Landnam (AD 985) to the Present (1985)', in Hilary H. Birks et al. (eds), *The Cultural Landscape*, Cambridge 1988, pp. 381–94.

68 Kirsten A. Seaver, *The Frozen Echo*, Stanford, CA 1996.

69 Gisli Gunnarson, *A Study of Causal Relations in Climate and History*, Lund 1980, pp. 13–15.

70 Gudrun Sveinbjarnardottir, *Farm Abandonment in Medieval and Post-Medieval Iceland: An Interdisciplinary Study*, Oxford 1992.

71 D. P. Willis, *Sand and Silence: Lost Villages of the North*, Aberdeen 1986.

72 Svend Gissel et al., *Desertion and Land Colonisation in the Northern Countries, c. 1300–1600*, Stockholm 1981.

73 Jean M. Grove, 'The Incidence of Landslides, Avalanches and Floods in Western Norway during the Little Ice Age', *Arctic and Alpine Research* 4 (1972), pp. 131–8.

74 K. J. Allison, *Deserted Villages*, London 1970.

75 Maurice W. Beresford, *The Lost Villages of England*, Gloucester 1987.

76 Trevor Rowley and John Wood, *Deserted Villages*, Princes Risborough 1995, pp. 16, 19.

77 Maurice W. Beresford and J. W. Hurst, *Wharram Percy: Deserted Medieval Village*, Princeton, NJ 1990.

78 Wilhelm Abel, *Die Wüstungen des ausgehenden Mittelalters*, 3rd edn, Stuttgart 1976.

79 Louis Gollut, *Mémoires historiques de la République séquanoise*, Dole 1592, lib. 2, cap. 18.

80 Fernand Braudel, *The Mediterranean and the Mediterranean World in the Age of Philip II*, vol. 1, London 1975, pp. 267ff.

81 William Chester Jordan, *The Great Famine*, Princeton, NJ 1996, pp. 7f.

82 David Arnold, *Famine, Social Crisis and Historical Change*, Oxford 1988.

83 Pierre Alexandre, *Le Climat en Europe au Moyen Âge*, Paris 1987, pp. 781–5.

84 Rüdiger Glaser, *Klimageschichte Mitteleuropas*, Darmstadt 2001, pp. 64f.

85 Henry S. Lucas, 'The Great European Famine of 1315, 1316, and 1317', *Speculum* 5 (1930), pp. 343–77.

86 Glaser, *Klimageschichte Mitteleuropas*, p. 65.

87 Jordan, *The Great Famine*, pp. 127–50.

88 Klaus Bergdolt, *Der schwarze Tod in Europa*, Munich 1994, p. 209.

89 Glaser, *Klimageschichte Mitteleuropas*, pp. 65f.

90 Trevor Rowley and John Wood, *Deserted Villages*, Princes Risborough 1995, p. 14.

91 Glaser, *Klimageschichte Mitteleuropas*, pp. 65f.

92 Bergdolt, *Der schwarze Tod in Europa*, pp. 208, 212.

93 J.-L. Biraben, *Les Hommes et la peste en France et dans les pays européens et méditerranéens*, 2 vols., Paris 1975.

94 David Herlihy, *The Black Death and the Transformation of the West*, Cambridge, MA 1997.

95 Neithard Bulst, 'Der Schwarze Tod: Demographische, wirtschafts- und kulturgeschichtliche Aspekte der Pestkatastrophe 1347–1352', *Saeculum* 30 (1979), pp. 45–67.

96 Heinrich Dormeier, 'Pestepidemien und Frömmigkeitsformen in Italien und Deutschland (14.-16. Jahrhundert)', in Manfred Jakubowski-Tiessen and Hartmut Lehmann (eds), *Um Himmels*

Willen: Religion in Katastrophenzeiten, Göttingen 2003, pp. 14–50.

97 Jordan, *The Great Famine*, pp. 185f.

98 Glaser, *Klimageschichte Mitteleuropas*, pp. 66f., 84f., 89.

99 Neidhart Bulst, 'Pest', in *Lexikon des Mittelalters* 6 (2003), esp. pp. 1915–18.

100 Moritz John Elsas, *Umrißeiner Geschichte der Preise und Löhne vom ausgehenden Mittelalter bis zum Beginn des 19. Jahrhunderts*, 2 vols., Leiden 1936/39.

101 Wilhelm Abel, *Agricultural Fluctuations in Europe: From the Thirteenth to the Twentieth Centuries*, London 1980.

102 Immanuel Wallerstein, *The Modern World System*, New York 1974.

103 Dietrich Saalfeld, 'Die Wandlungen der Preis- und Lohnstruktur während des 16. Jahrhunderts in Deutschland', in Wolfram Fischer (ed.), *Beiträge zu Wirtschaftswachstum und Wirtschaftsstruktur im 16. und 19. Jahrhundert*, Berlin 1971, pp. 9–28.

104 Edward A. Eckert, *The Structure of Plague and Pestilences in Early Modern Europe*, Basel 1996.

105 Carlo M. Cipolla, *Christophano and the Plague*, London 1973.

106 Eckert, *The Structure of Plague and Pestilences in Early Modern Europe*, p. 150.

107 John D. Post, 'Famine, Mortality and Epidemic Disease in the Process of Modernization', *Economic History Review* 29 (1976), pp. 14–37. Cf. Andrew B. Appleby, 'Famine, Mortality and Epidemic Disease: A Comment', *Economic History Review* 30 (1977), pp. 508–12.

108 Massimo Livi-Bacci, *Population and Nutrition*, Cambridge 1991.

109 S. R. Duncan, Susan Scott and C. J. Duncan, 'The Dynamics of Smallpox Epidemics in Britain, 1550–1800', *Demography* 30 (1993), pp. 405–23.

110 Jean Dubos, *The White Plague: Tuberculosis, Man and Society*, Boston 1952.

111 Andrew B. Appleby, 'Disease or Famine? Mortality in Cumberland and Westmoreland, 1580–1640', *Economic History Review* 26 (1973), pp. 403–32.

112 Ernst Woehlkens, *Pest und Ruhr im 16. und 17. Jahrhundert*, Uelzen 1954.

113 Emmanuel Le Roy Ladurie, 'L'Aménhorrée de famine (XVIIe–XXe siècles)', *Annales d'histoire économique et sociale* 24 (1969), pp. 1589–97.
114 Helmut Wurm, 'Körpergröße und Ernährung der Deutschen im Mittelalter', in Bernd Hermann (ed.), *Mensch und Umwelt im Mittelalter*, Stuttgart 1986, pp. 101–8.
115 Christoph Schorer, *Memminger Chronik*, Memmingen 1660, p. 101.
116 Wolfgang Behringer, 'Die Hungerkrise von 1570: Ein Beitrag zur Krisengeschichte der Neuzeit', in Manfred Jakubowski-Tiessen and Hartmut Lehmann (eds), *Um Himmels Willen: Religion in Katastrophenzeiten*, Göttingen 2003, pp. 51–156.
117 Frederic Wakeman, 'China and the Seventeenth-Century Crisis', *Late Imperial China* 7 (1986), pp. 1–26.
118 Anthony Reid, 'The Seventeenth Century Crisis in South East Asia', *Modern Asian Studies* 24 (1990), pp. 639–59.
119 *Das Gesamtwerk von Brueghel*, with an introduction by Charles de Tolnay and a scholarly appendix by Piero Bianconi, Zurich 1967, plates X–XIII.
120 Wolfgang Behringer, 'Mörder, Diebe, Ehebrecher: Verbrechen und Strafen in Kurbayern vom 16. bis 18. Jahrhundert', in Richard van Dülmen (ed.), *Verbrechen, Strafen und soziale Kontrolle. Studien zur historischen Kulturforschung*, vol. 3, Frankfurt/Main 1990, pp. 85–132, 287–93.
121 Edward Peters, *Torture*, Philadelphia PA 1996.
122 Richard van Dülmen, *Theatre of Horror: Crime and Punishment in Early Modern Germany*, Cambridge 1990.
123 Günter Franz, *Der Dreißigjährige Krieg und das deutsche Volk*, Leipzig 1940.
124 Wolfgang Behringer, 'Von Krieg zu Krieg: Neue Perspektiven auf das Buch von Günther Franz *Der Dreißigjährige Krieg und das deutsche Volk* (1940)', in Benigna von Krusenstern and Hans Medick (ed.), *Zwischen Alltag und Katastrophe: Der Dreißigjährige Krieg aus der Nähe*, Göttingen 1999, pp. 543–91.
125 Edward A. Eckert, *The Structure of Plagues and Pestilences in Early Modern Europe*, Basel 1996, p. 150.
126 Kenneth J. Hsü, *Klima macht Geschichte*, Zurich 2000, pp. 33ff.
127 'Ettlich Hundert Herrlicher und schönner Carmina oder gedicht/von der Lanngwürigen schweren gewesten Theuerung/

grossen Hungers Not/und allerlay zuvor unerhörten Grausamen Straffen/und Plagen/so wir (Gott Lob) zum tail ausgestanden haben [. . .]', in Stadtarchiv Augsburg, *Memorbuch Paul Hektor Mairs*, fol. 800–34.

128 Pitirim Sorokin, *Man and Society in Calamity: The Effects of War, Revolution, Famine, Pestilence upon Human Mind, Behavior, Social Organization and Cultural Life*, New York 1942.

129 R. W. Perry, 'Environmental Hazards and Psychopathology: Linking Natural Disasters with Mental Health', *Environmental Management* 7 (1983), pp. 331–9.

130 R. Jay Turner, Blair Wheaton and Donald A. Lloyd, 'The Epidemiology of Social Stress', *American Sociological Review* 60 (1995), pp. 104–25.

131 Robert Burton, *The Anatomy of Melancholy*, Oxford 1621.

132 Norman Rosenthal, *Winter Blues*, New York 1993.

133 See the home page of the SAD Association: www.sada.org.uk/sadassociation.htm.

134 Raymond W. Lam, 'Major Depressive Disorder: Seasonal Affective Disorder', *Current Opinion in Psychiatry*, January 1994.

135 Jelle Zeilinga de Boer and Donald Theodore Sanders, *Volcanoes in Human History: The Far-Reaching Effects of Major Eruptions*, Princeton, NJ 2002.

136 Vilhjalmar Bjarnar, 'The Laki Eruption [1783/84] and the Famine of the Mist', in Carl F. Bayerschmidt and Erik J. Friis (eds), *Scandinavian Studies*, Seattle 1965, pp. 410–21.

137 Michel de Montaigne, *Essays*, trans. Charles Cotton, London 1877, Chapter 2, 'Of Sorrow'.

138 Simon Musaeus, *Melancholischer Teufel nützlicher Bericht/wie man alle Melancholische Teufflische gedancken von sich treiben soll*, Tham in der Neumark 1572.

139 Daniel Schaller, *Herold*, Magdeburg 1595, pp. 129f.

140 David Lederer, 'Verzweiflung im Alten Reich: Selbstmord während der "Kleinen Eiszeit"', in Wolfgang Behringer et al. (eds), *Kulturelle Konsequenzen der 'Kleinen Eiszeit'*, Göttingen 2005, pp. 255–82. Cf. Lederer, *Madness, Religion and the State in Early Modern Europe: A Bavarian Beacon*, Cambridge 2006.

141 David Lederer, 'Aufruhr auf dem Friedhof: Pfarrer, Gemeinde und Selbstmord im frühneuzeitlichen Bayern', in Gabriela

Signori (ed.), *Trauer, Verzweiflung und Anfechtung*, Tübingen 1994, pp. 189–209, pp. 201f.

142 Andrés Velásquez, *Libro de la Melancholia*, s.l. 1585; André de Laurens, *Des Maladies mélancoliques*, Paris 1597; Alonso de Santa Cruz, *De Melancholia*, s.l. 1613.

143 Lawrence Babb, *The Elizabethan Malady*, East Lansing, MI 1951.

144 Michael Macdonald, *Sleepless Souls: Suicide in Early Modern England*, Oxford 1990.

145 Paul S. Seaver, *Wallington's World: A Puritan Artisan in Seventeenth Century London*, Stanford, CA 1985.

146 Richard J. W. Weston, *Rudolf II and His World*, Oxford 1973.

147 Gertrude von Schwarzenfeld, *Rudolf II.: Der saturnische Kaiser*, Munich 1961.

148 Felix Stieve, *Die Verhandlungen über die Nachfolge Kaiser Rudolf II. in den Jahren 1581–1602*, Munich 1879, p. 48.

149 Erik Midelfort, *Mad Princes of Renaissance Germany*, Charlottesville, VA 1994, pp. 132–70.

150 Ibid., p. 178.

151 Raymond Klibansky, Erwin Panofsky and Fritz Saxl, *Saturn and Melancholy. Studies in the History of Natural Philosophy, Religion and Art*, London 1964.

152 H. C. Erik Midelfort, *A History of Madness in Sixteenth-Century Germany*, Stanford, CA 1999, p. 162; summarizing Andreas Planer [praes.] and Johann Faber [resp.], *De morbo Saturnino seu melancholia*, Tübingen 1593.

153 Johann Weyer, *De Praestigiis Daemonum*, Basel 1563; reissued Frankfurt/Main 1986. Extracted in Wolfgang Behringer (ed.), *Hexen und Hexenprozesse in Deutschland*, 5th edn, Munich 2001, p. 144.

154 Richard L. Kagan, *Lucrecia's Dreams: Politics and Prophecy in Sixteenth-Century Spain*, Berkeley 1990.

4 CULTURAL CONSEQUENCES OF THE LITTLE ICE AGE

1 'Von einer grusamen und erschrohenlicher gesicht, die am himmel wyt und breyt gesähen am 28. Decemb. dess verschinen 1560. iars, 1561', in Matthias Senn (ed.), *Die Wickiana*, Zurich 1975, p. 55.

2 'Von dem grossen fhürigen zeichen, welches an unschuldigen kindlinen tag [28.12.1560] gesähen an vilen Orten, 1561', in ibid., p. 58.

3 'Uff den fhürigen Himmel ist ein unsagliche grose kelte gefolget, 1561', in ibid., pp. 58f.

4 Manfred Jakubowski-Tiessen, 'Das Leiden Christi und das Leiden der Welt: Die Entstehung des lutherischen Karfreitags', in Wolfgang Behringer et al. (eds), *Kulturelle Konsequenzen der 'Kleinen Eiszeit'*, Göttingen 2005, pp. 195–214.

5 Carlos M. Eire, *From Madrid to Purgatory*, Cambridge 1995.

6 For the years 1564–6 alone we find dozens of such consolatory writings in the book fair catalogues. Two examples: Nikolaus Selnecker, *Bericht vom sterben / und von sterbenden leuffen / und wie man sich Christlich darinn halten und trösten soll*, Leipzig 1565; Johann Lang, *Trostbüchlein: Wie man die kranken und sterbenden besuchen und trösten soll*, Lauingen 1566.

7 Leonhard Lenk, *Augsburger Bürgertum im Späthumanismus und Frühbarock*, Augsburg 1968, pp. 82–6.

8 Andreas Gryphius, *Gesamtausgabe der deutschsprachigen Werke*, 8 vols., Tübingen 1963–72.

9 Heinz Schilling, 'Geschichte der Sünde oder Geschichte des Verbrechen?', *Annali dell'istituto storico italo-germanico in Trento* 12 (1986), pp. 169–92.

10 William Monter, 'Sodomy and Heresy in Early Modern Switzerland', *Journal of Homosexuality* 8 (1980–1), pp. 41–53.

11 Richard van Dülmen, *Theatre of Horror: Crime and Punishment in Early Modern Germany*, Cambridge 1990.

12 Günter Pallaver, *Das Ende der schamlosen Zeit: Die Verdrängung der Sexualität in der frühen Neuzeit am Beispiel Tirols*, Vienna 1987.

13 Heinrich Richard Schmidt, *Dorf und Religion*, Stuttgart 1995.

14 *Das Buch Weinsberg: Kölner Denkwürdigkeiten aus dem 16. Jahrhundert*, vol. 5, Bonn 1926, pp. 193f.

15 Howard S. Becker, *Outsiders*, New York 1963, pp. 147–63; here, p. 148.

16 Michel de Montaigne, *Essays*, trans. Charles Cotton, London 1877, Chapter 11, 'Of Cripples'.

17 Will-Erich Peuckert, 'Religiöse Unruhe um 1600', in Richard Alewyn (ed.), *Deutsche Barockforschung: Dokumentation einer Epoche*, Cologne, Berlin 1968, pp. 75–93.

18 Winfried Schulze, 'Untertanenrevolten, Hexenverfolgungen und "Kleine Eiszeit"? Eine Krise um 1600', in Bernd Roeck,

Klaus Bergdolt and Andrew John Martin (eds), *Venedig und Oberdeutschland in der Renaissance: Beziehungen zwischen Kunst und Wirtschaft*, Sigmaringen 1993, pp. 289–312.

19 'Judenfeindschaft', in *Lexikon des Mittelalters* 5 (1999), columns 790–2.

20 Friedrich Battenberg, *Das europäische Zeitalter der Juden*, vol. 1, Darmstadt 1990, pp. 91f.

21 Malcolm Barber, 'Lepers, Jews and Moslems: The Plot to Overthrow Christendom in 1321', *History* 66 (1981), pp. 1–17.

22 František Graus, *Pest – Geißler – Judenmorde*, Göttingen 1987, pp. 155–274, 299–334.

23 Battenberg, *Das europäische Zeitalter der Juden*, vol. 1, Darmstadt 1990, pp. 123f.

24 Norman Cohn, *Europe's Inner Demons: An Enquiry Inspired by the Great Witch-Hunt*, London 1975.

25 Wolfgang Behringer, 'Climatic Change and Witch-Hunting: The Impact of the Little Ice Age on Mentalities', *Climatic Change* 43 (1999), pp. 335–51.

26 Richard Golden (ed.), *The Encyclopedia of Witchcraft*, 4 vols., Santa Barbara, CA 2006.

27 Johannes Franck, 'Geschichte des Wortes Hexe', in Joseph Hansen (ed.), *Quellen und Untersuchungen zur Geschichte des Hexenwahns und der Hexenverfolgung im Mittelalter*, Bonn 1901, pp. 614–70.

28 Wolfgang Behringer, 'Weather, Hunger and Fear: The Origins of the European Witch Persecution in Climate, Society and Mentality', *German History* 13 (1995), pp. 1–27.

29 Wolfgang Behringer, 'Neun Millionen Hexen: Entstehung, Tradition und Kritik eines populären Mythos', *Geschichte in Wissenschaft und Unterricht* 49 (1998), pp. 664–85.

30 Günter Jerouschek and Wolfgang Behringer (eds), Heinrich Kramer (Institoris), *Der Hexenhammer: Malleus maleficarum*, trans. from the Latin by Günter Jerouschek and Wolfgang Behringer (eds) and Werner Tschacher, Munich 2000.

31 Johann Weyer, 'Vorwort' and 'Anhang', *De Praestigiis Daemonum*, Basel 1563.

32 Wolfgang Behringer, *Witches and Witch-Hunts: A Global History*, Cambridge 2004.

33 Caspar Macer, *Drei Bittpredigten: 1. Von der großen Theuerung, 2. Vom Krieg und Blutvergießen, 3. Von der Pestilenz, gehalten zu Regensburg*, Munich 1572.

34 Otto Ulbricht, 'Extreme Wetterlagen im Diarium Heinrich Bullingers (1504–1574)', in Wolfgang Behringer et al. (eds), *Kulturelle Konsequenzen der 'Kleinen Eiszeit'*, Göttingen 2005, pp. 149–78.

35 Jacobus Feucht, *Fünf Predigten zur Zeit der grossen Theuerung / Hungersnoth und Ungewitter / darinn die fünff fürnembsten ursachen des Göttlichen Zorns angezeigt werden*, Cologne 1574.

36 Ludwig Lavater, *Von thüwre und hunger, dry Predigen [. . .]*, Zurich 1571.

37 Thomas Rorarius, *Fuenff und zwentzig Nothwendige Predigten von der Grausamen regierenden Thewerung*, Frankfurt/ Main 1572. See also *Zwo Predig, wie man sich Christlich halten soll, wann grosse Ungewitter oder Hagel sich erheben, mit [. . .] Underrichtung von dem Leutten gegen Wetter [. . .]. Die erst D. Johannes Brentzen. Die ander Thoman Roerers. Das dritt M. Christoffen Vischers*, Nuremberg 1570.

38 *Predigt über Hunger- und Sterbejahre, von einem Diener am Wort*, s.l. 1571.

39 *Neue Konkordanz zur Einheitsübersetzung der Bibel*, ed. by Franz Joseph Schierse, retranslated by Winfried Bader, Darmstadt 1996, *Hagel*, pp. 632f., *Hunger*, pp. 782ff., *Pest*, p. 1241, *Zorn Gottes*, pp. 2024–31.

40 Gordon Manley, 'Climatic Fluctuations and Fuel Requirements', *Scottish Geographical Magazine* 73 (1957), pp. 19–28.

41 *Das Buch Weinsberg: Kölner Denkwürdigkeiten aus dem 16. Jahrhundert*, vol. 5, Bonn 1926, p. 354.

42 Daniel Schaller, *Herold*, Magdeburg 1595, pp. 156f.

43 Richard van Dülmen, *Kultur und Alltag in der Frühen Neuzeit*, vol. 1, Munich 1990, pp. 56–68.

44 Bernd Roeck, *Lebenswelt und Kultur des Bürgertums in der Frühen Neuzeit*, Munich 1991, pp. 15f.

45 Michel de Montaigne, *The Journal of Montaigne's Travels*, tr. by W.G. Watters, London 1903.

46 *Das Buch Weinsberg*, vol. 5, p. 213.

47 Ibid., vol. 3, Bonn 1897, pp. 256–8.

48 Ernst Walter Zeeden, *Deutsche Kultur in der Frühen Neuzeit*, Frankfurt/Main 1968, p. 308.

49 *Das Buch Weinsberg*, vol. 5, pp. 121 (sleeping costume), 213 (new sleeping costume), 360f. (eating).

50 Andreas Musculus, *Vermahnung und Warnung vom zerluder-ten, zucht- und ehrverwegenen pludrigten Hosenteufel*, Leipzig 1555.

51 *Theatrum Diabolorum*, Frankfurt/Main 1569, pp. 388ff.

52 Johann Strauß, *Wider den Kleider/Pluder/Pauß/und Krauß-Teuffel*, Freiberg 1581; Lukas Osiander, *Ein Predig von Hoffertigen/ungestalter Kleydung der Weibs und Mannspersonen*, Tübingen 1586.

53 Ludmila Kybalova et al., *Das große Bilderlexikon der Mode*, Gütersloh, Berlin 1975, pp. 139–62.

54 *Das Buch Weinsberg*, vol. 5, pp. 256f., 269.

55 John Vanderbank, *Francis Bacon*, National Portrait Gallery, London.

56 *Das Buch Weinsberg*, vol. 2, p. 377.

57 R. an der Heiden, 'Die Porträtmalerei des Hans von Aachen', *Jahrbuch der Kunsthistorischen Sammlungen* 66 (1970), pp. 135–226.

58 Kybalova et al., *Das große Bilderlexikon der Mode*, pp. 163–89.

59 *Prag um 1600: Kunst und Kultur am Hofe Kaiser Rudolfs II.*, 2 vols., Freren/Emsland 1988.

60 J. Anderson Black and Madge Garland, *A History of Fashion*, London 1975, pp. 165f.

61 *Das Buch Weinsberg*, vol. 3, p. 257.

62 Johann Reinhold, *Predig wider den unbändigen Putzteufel*, Frankfurt/Main 1609, p. 3.

63 Aegidius Albertinus, *Luzifers Königreich und Seelengejaid*, Munich 1616, pp. 106f.

64 Thomas Da Costa Kaufmann, 'Arcimboldo and Propertius: A Classical Source for Rudolf II as Vertumnus', *Zeitschrift für Kunstgeschichte* 48 (1985), pp. 117–23.

65 Arnold Hauser, *The Social History of Art*, vol. 2, *Renaissance, Mannerism, Baroque*, 2nd edn, London 1962.

66 Andreas Tönnesmann, *Der europäische Manierismus 1520–1610*, Munich 1997.

67 Hans Neuberger, 'Climate in Art', *Weather* 25 (1970), No. 2, pp. 46–56.

68 Johann Georg Prinz zu Hohenzollern (ed.), *Von Greco bis Goya*, Munich 1982, pp. 52f., 156–67.

69 *Das Gesamtwerk von Bruegel*, Zurich 1967, p. 6 and plates XXXII and XXXIII.

70   F. Groissmann, *Pieter Bruegel*, 3rd rev. edn, London, New York 1973.

71   *Das Gesamtwerk von Bruegel*, p. 6 and plates XXVIII to XXXI.

72   Ibid., plates XL to XLIII and XLVII.

73   Lawrence Otto Goedde, *Tempest and Shipwreck in Dutch and Flemish Art*, London 1989.

74   Knut Frydendahl and H. H. Lamb, *Historic Storms of the North Sea*, Cambridge 1991.

75   Hans Khevenhüller, *Geheimes Tagebuch, 1548–1605*, Graz 1971, p. 89.

76   Orlando di Lasso, *Bußpsalmen*, Munich 1572.

77   Patrice Veit, ' "Gerechter Gott, wo will es hin / Mit diesen kalten Zeiten?" Witterung, Not und Frömmigkeit im evangelischen Kirchenlied', in Wolfgang Behringer et al. (eds), *Kulturelle Konsequenzen der 'Kleinen Eiszeit'*, Göttingen 2005, pp. 283–310.

78   Paul Münch, *Das Jahrhundert des Zwiespalts*, Stuttgart 1999, pp. 139ff.

79   Karl G. Fellerer, *Der Stilwandel in der europäischen Musik um 1600*, Opladen 1972.

80   Johannes Janssen, *Geschichte des deutschen Volkes seit dem Ausgang des Mittelalters*, vol. 6, *Kunst und Volksliteratur bis zum Beginn des dreißigjährigen Krieges*, Freiburg/Breisgau 1893, pp. 425–57.

81   Bernd Roeck, 'Renaissance – Manierismus – Barock: Sozial- und klimageschichtliche Hintergründe künstlerischer Stilveränderungen', in Wolfgang Behringer et al. (eds), *Kulturelle Konsequenzen der 'Kleinen Eiszeit'*, Göttingen 2005, pp. 323–47.

82   Eliška Fučíková, 'The Collection of Rudolf II at Prague: Cabinet of Curiosities or Scientific Museum?', in Oliver Impey and Arthur MacGregor (eds), *The Origins of the Museum*, Oxford 1986, pp. 49–53.

83   Sigmund Feyerabend (ed.), *Theatrum Diabolorum*, Frankfurt/Main 1569.

84   Peter Schmidt (ed.), *Erster und ander Theil Theatri Diabolorum*, Frankfurt/Main 1587.

85   Abraham Sawr (ed.), *Theatrum de Veneficis*, Frankfurt/Main 1586.

86   James VI of Scotland, *Daemonologie, in forme of a dialoge*, Edinburgh 1597.

87 Stuart Clark, *Thinking with Demons*, Oxford 1999.
88 Hans Rupprich, *Die deutsche Literatur vom Spätmittelalter zum Barock*, vol. 2, Munich 1973, pp. 191–8.
89 Gustav René Hocke, *Manierismus in der Literatur*, Reinbek 1959.
90 Richard Newald, *Die deutsche Literatur vom Späthumanismus zur Empfindsamkeit 1570–1750*, Munich 1967, pp. 18f.
91 Andreas Gryphius, *Gesamtausgabe der deutschsprachigen Werke*, vol. 1, Tübingen 1963, p. 33.
92 Marian Szyrocki, *Die deutsche Literatur des Barock*, Stuttgart 1979, p. 147.
93 Klaus Garber, *Der locus amoenus und der locus terribilis*, Cologne, Vienna 1974.
94 Martin Opitz, *Buch von der deutschen Poeterey* [Breslau 1624], Stuttgart 1970.
95 Marian Szyrocki, *Die deutsche Literatur des Barock*, Stuttgart 1979, pp. 170ff.
96 Wolfram Mauser, 'Was ist dies Leben doch? Zum Sonett "Thränen in schwerer Kranckheit" von Andreas Gryphius', in Volker Meid (ed.), *Gedichte und Interpretationen*, vol. 1, Stuttgart 1982, pp. 222–30.
97 Ferdinand von Ingen, *Vanitas und Memento Mori in der deutschen Barocklyrik*, Groningen 1966.
98 Susan Reynolds Whyte, *Questioning Misfortune: The Pragmatics of Uncertainty in Eastern Uganda*, Cambridge 1997.
99 Charles Tilly (ed.), *The Formation of National States in Western Europe*, Princeton, NJ 1975.
100 Jan de Vries, 'Analysis of Historical Climate–Society Interactions', in Robert W. Kates, Jesse H. Ausubel and Mimi Berberian (eds), *Climate Impact Assessment*, Chichester 1985, pp. 273–92, 286f.
101 Theodore K. Rabb, *The Struggle for Stability in Early Modern Europe*, New York 1975.
102 Immanuel Wallerstein, *The Modern World-System*, New York 1974.
103 Thomas Hobbes, *Leviathan*, London 1651.
104 Thomas Robisheaux, *Rural Society and the Search for Order in Early Modern Germany*, Cambridge 1989.
105 Markus Raeff, *The Well-Ordered Police State*, New Haven, CT 1983.
106 Michael Stolleis, *Staat und Staatsräson in der frühen Neuzeit*, Frankfurt/Main 1990.

107 Norbert Elias, *The Civilizing Process*, Oxford 1994.
108 'Post-Ordnung', in Johann Heinrich Zedler (ed.), *Großes vollständiges Universal-Lexicon aller Wissenschaften und Künste*, 64 vols., Halle/Leipzig 1732–54, vol. 28 (1741), columns 1812–27.
109 Johannes Kepler, *The Harmony of the World*, Philadelphia, PA 1997 [Johannes Kepler, Harmonices mundi, Frankfurt/Main 1619].
110 Henning Eichberg, 'Geometrie als barocke Verhaltensnorm, Fortifikation und Exerzitien', *Zeitschrift für Historische Forschung* 4 (1977), pp. 17–50.
111 Marian Szyrocki (ed.), *Poetik des Barock*, Stuttgart 1977.
112 Willer, *Herbstmesse 1570*; Fabian edition, s.l. 1972, I, pp. 308–11.
113 Konrad Heresbach, *Rei rusticae libri quatuor. Vier Bücher über Landwirtschaft* [1570], ed. by Wilhelm Abel, Meisenheim 1970.
114 M. C. Heresbachius [Konrad Heresbach], *Foure Bookes of Husbandry*, London 1577.
115 Charles Estienne, *L'Agriculture & Maison rustique*, Paris 1572 [first edn 1564, reprinted 1567; later edns 1576, 1583, 1589, 1598, 1602, 1625, 1653, 1677]; German translation 1579; English translation: *Maison Rustique, or the Countrie Farme*, London 1600.
116 Martin Grosser, *Kurtze und einfeltige anleytung Zu der Landtwirtschafft; beyder im Ackerbaw, und in der Viehzucht*, s.l. 1590; Johannes Colerus, *Oeconomia ruralis et domestica*, Wittenberg 1593 [German translation 1593].
117 Klaus Herrmann, *Pflügen, Säen, Ernten: Landarbeit und Landtechnik in der Geschichte*, Reinbek 1985, p. 112.
118 Bartholomäus Scultetus, *Ein ewigwerend Prognosticon / von aller Witterung in der Lufft / und der Wercken der andern Element*, Görlitz 1572.
119 Tobias Lotter, *Gründlicher und nothendiger Bericht, was von denen ungestümen Wettern, verderblichen Hägeln und schädlichen Wasserfluten, mit welchen Teutschland an sehr vielen orten in dem 1613. Jar ernstlich heimgesucht worden, zuhalten seye [. . .]*, Stuttgart 1615.
120 Johann Georg Sigwart, *Ein Predigt Vom Reiffen und Gefröst, den 25. Aprilis [. . .] 1602 (als die nächste Tag zuvor, nemblich den 21., 22., und 23. gemelten Mondts das Rebwerck erfroren), [. . .]*, Tübingen 1602; and idem, *Ein Predigt Vom Hagel und*

Ungewitter, Im Jahr Christi 1613, den 30 May [. . .] als am Sambstag Abends zuvor Nachmittag vor 5 Uhren ein schröcklicher Hagel gefallen [. . .],Tübingen 1613.

121 Wolfgang Behringer, 'Climatic Change and Witch-Hunting: The Impact of the Little Ice Age on Mentalities', *Climatic Change* 43 (1999), pp. 335–51.

122 Wolfgang Behringer, *Im Zeichen des Merkur: Reichspost und Kommunikationsrevolution in der Frühen Neuzeit*, Göttingen 2003.

123 Günter Abel, *Stoizismus und Frühe Neuzeit*, Berlin 1978.

124 Galileo Galilei, *Dialogo sopra i due massimi sistemi*, Florence 1632; *Dialogue on the Great World Systems*, the Salusbury translation of 1661 revised by Giorgio de Santillana, Chicago 1953.

125 Francis Bacon, *Novum Organon Scientiarum* [1620]; *The New Organon*, Cambridge 2000.

126 Eveline Cruikshanks, *The Glorious Revolution*, London 2000.

127 Lynne Thorndike, *A History of Magic and Experimental Science*, 8 vols., New York 1923–58.

128 Benvenuto Cellini, *The Life of Benvenuto Cellini*, London 1960.

129 Girolamo Cardano, *Des Girolamo Cardano von Mailand eigene Lebensbeschreibung*, Kempten 1969.

130 Gianbattista della Porta, *Magia naturalis in libri XX*, Naples 1589; English translation of 1658, *Natural Magick*, available at http://homepages.tscnet.com/omard1/jportat3.html.

131 Peter J. French, *John Dee: The World of an Elizabethan Magus*, New York 1989.

132 Bruce T. Moran, *The Alchemical World of the German Court: Occult Philosophy and Chemical Medicine in the Circle of Moritz of Hessen (1572–1632)*, Stuttgart 1991.

133 Isaac Newton, *Philosophiae Naturalis Principia Mathematica*, London 1687.

134 Herbert Butterfield, *The Origins of Modern Science*, London 1965.

135 E. J. Dijksterhuis, *The Mechanization of the World Picture*, Oxford 1961.

136 Thomas S. Kuhn, *The Structure of Scientific Revolutions*, Chicago 1962.

137 Roy Porter, *The Creation of the Modern World*, New York 2000, p. 149.

138 Samuel E. Finer, 'State- and Nation Building in Europe', in Tilly (ed.), *The Formation of National States in Western Europe*, pp. 84–163.

139 Peter Burke, *The Fabrication of Louis XIV*, London 1992.

140 Burkhard Frenzel (ed.), *Climatic Trends and Anomalies in Europe 1675–1715*, Stuttgart 1994.

141 Marcel Lachiver, *Les Années de misère: la famine au temps du Grand Roi, 1680–1720*, Paris 1991.

142 S. Lindgren and J. Neumann, 'The Cold Wet Year 1695', *Climatic Change* 3 (1981), pp. 173–87.

143 Eeno Jutikkala, 'The Great Finnish Famine in 1696/97', *Scandinavian Economic History Review* 3 (1955), pp. 48–63.

144 Patrice Berger, 'French Administration in the Famine of 1693', *European Studies Review* 8 (1978), pp. 101–27.

145 W. Gregory Monahan, *Years of Sorrows: The Great Famine of 1709 in Lyon*, Columbus, OH 1993.

146 Jonathan Israel, *The Dutch Republic: Its Rise, Greatness and Fall 1477–1806*, Oxford 1995, pp. 334f.

147 Klaus Herrmann, *Pflügen, Säen, Ernten*, Reinbek 1985, pp. 115–38.

148 Israel, *The Dutch Republic*, Oxford 1995.

149 Michael Budde (ed.), *Die 'Kleine Eiszeit'*, Berlin 2001.

150 Paul Hazard, *European Thought in the Eighteenth Century*, Harmondsworth 1965.

151 Mark Overton, *Agricultural Revolution in England*, Cambridge 1996.

152 L. C. Madonna, *Christian Wolff und das System des klassischen Rationalismus*, Hildesheim 2001.

153 *Kurtze zufällige und vermischte Gedancken, über den hefftigen Schnee und Frost-Winter*, Tübingen 1740.

154 *Observationes Meteorologicae*, Weißenburg am Nordgau 1740, pp. 42f.

155 Johann Rudolph Marcus, *Nachricht von dem im ietzigen 1740ten Jahre eingefallenen ausserordentlich strengen und langen Winter*, Leipzig n.d. [1740].

156 M. G. Pearson, 'The Winter of 1739–40 in Scotland', *Weather* 28 (1973), pp. 20–4.

157 John Barker, *An Inquiry into the Nature, Cause and Cure of the Present Epidemic Fever*, London 1742.

158 Marcus, *Nachricht . . .*, pp. 15f.

159 Gordon Manley, 'The Great Winter of 1740', *Weather* 13 (1958), pp. 11–17; K. J. Gruffyd, 'The Vale of Clwyd Corn

Riots of 1740', *Flintshire Publications* 27 (1975–6), pp. 36–42; B. W. Alexander, 'The Epidemic Fever (1741–42)', *Salisbury Medical Bulletin* 11 (1971), pp. 24–9; Michael Drake, 'The Irish Demographic Crisis of 1740–41', *Historical Studies* 6 (1968), pp. 101–24.

160 M. Deutsch et al., *Der Winter 1739/40 in Halle/Saale*, Berlin 1996.

161 John D. Post, *Food Shortage, Climatic Variability, and Epidemic Disease in Preindustrial Europe: The Mortality Peak in the Early 1740s*, London 1985.

162 H. Arakawa, 'Meteorological Conditions of the Great Famines in the Last Half of the Tokugawa Period, Japan', *Papers in Meteorology and Geophysics* (Tsukuba) 6 (1955), pp. 101–15.

163 Kiyoshi Inoue, *Geschichte Japans* [1963], Frankfurt/Main 1993, pp. 261ff., 266f.

164 Vilhjalmar Bjarnar, 'The Laki Eruption and the Famine of the Mist', in Carl F. Bayerschmidt and Erik J. Friis (eds), *Scandinavian Studies*, Seattle 1965, pp. 410–21.

165 Sigurd Thorarinsson, 'The Lakagigar Eruption of 1783', *Bulletin Volcanologique* 33 (1969), pp. 910–29.

166 Gaston R. Demarée et al., 'Bons Baisers d'Islande: Climatic, Environmental and Human Dimensions Impacts of the Lakigigar Eruption (1783–1784) in Iceland', in Phil D. Jones et al. (eds), *History and Climate: Memories of the Future?*, Dordrecht 2001, pp. 219–46.

167 Johann Ernst Basilius Wiedeburg, 'Über die Erdbeben und den allgemeinen Nebel von 1783', *Göttingische Anzeigen von gelehrten Sachen* 47 (1784), pp. 470–2.

168 Benjamin Franklin, 'Meteorological Imaginations and Conjectures', *Memoirs of the Manchester Literary and Philosophical Society* 2 (1785), pp. 373–7.

169 Manfred Vasold, 'Die Eruptionen des Laki von 1783/84: Ein Beitrag zur deutschen Klimageschichte', *Naturwissenschaftliche Rundschau* 57 (2004), pp. 602–8.

170 Alan Taylor, '"The Hungry Year": 1789 on the Northern Border of Revolutionary America', in Alessa Johns (ed.), *Dreadful Visitations: Confronting Natural Catastrophe in the Age of Enlightenment*, New York 1999, pp. 145–81.

171 Jack A. Goldstone, *Revolution and Rebellion in the Early Modern World*, Berkeley 1991.

172 Chris E. Paschold and Albert Gier (eds), *Die Französische Revolution*, Stuttgart 1989, pp. 47f.

173 Georges Lefebvre, *The Great Fear of 1789*, London 1973.

174 Eberhard Weis, *Der Durchbruch des Bürgertums 1776–1847*, Berlin 1982, p. 300.

175 Eberhard Weis, 'Frankreich von 1661 bis 1789', in Theodor Schieder (ed.), *Handbuch der Europäischen Geschichte*, vol. 4, Stuttgart 1968, pp. 166–307: here p. 270.

176 Tom Simkin et al. (eds), *Volcanoes of the World*, Stroudsberg, PA 1981, pp. 112–31.

177 Henry Stommel and Elizabeth Stommel, *Volcano Weather*, Newport, RI 1983.

178 C. R. Harrington (ed.), *The Year without a Summer? World Climate in 1816*, Ottawa 1992.

179 Wolfgang Behringer, *Witches and Witch-Hunts: A Global History*, Cambridge 2004.

180 Jacob Katz, *Die Hep-Hep Verfolgungen des Jahres 1819*, Berlin 1994.

181 Jörn Sieglerschmidt, 'Untersuchungen zur Teuerung in Südwestdeutschland 1816/17', in *Festschrift Hans-Christoph Rublack*, Frankfurt/Main 1992, pp. 113–44.

182 Gerald Müller, *Hunger in Bayern, 1816–1818: Politik und Gesellschaft in einer Staatskrise des frühen 19. Jahrhunderts*, Frankfurt/Main 1998.

183 James Jameson, *Report on the Epidemic Cholera Morbus as It Visited the Territories Subject to the Presidency of Bengal in the Years 1817, 1818 and 1819*, Calcutta 1820.

184 Richard Evans, *Death in Hamburg: Society and Politics in the Cholera Years, 1830–1910*, Oxford 1987.

185 Wolfgang U. Eckart, 'Cholera', in *Enzyklopädie der Neuzeit 2* (2005), columns 717–20.

186 Cecil-Blanche Woodram-Smith, *Great Hunger: Ireland 1845–1849*, London 1962.

187 Joachim Schaier, *Verwaltungshandeln in einer Hungerkrise: Die Hungersnot 1846/47 im badischen Odenwald*, Wiesbaden 1988.

188 Thomas Martin Devine and Willie Orr, *The Great Highland Famine*, Edinburgh 1988.

189 Michael Maurer, *Kleine Geschichte Irlands*, Stuttgart 1998, pp. 219–26.

190 Christine Kinealy, *A Death-dealing Famine: The Great Hunger in Ireland*, London 1997.

191 H. Arakawa, 'Meteorological Conditions of the Great Famines in the Last Half of the Tokugawa Period, Japan',

Papers in Meteorology and Geophysics 6 (1955), Tsukuba, pp. 112ff.

192 Kiyoshi Inoue, *Geschichte Japans* [1963], Frankfurt/Main 1993, pp. 309ff.

5 GLOBAL WARMING: THE MODERN WARM PERIOD

1 Josef Ehmer, 'Bevölkerung', in *Enzyklopädie der Neuzeit* 2 (2005), columns 94–119.

2 David Blackbourn, *The Conquest of Nature*, London 2006.

3 Frank Konersmann, 'Agrarrevolution', in *Enzyklopädie der Neuzeit* 2 (2005), columns 131–6.

4 Mary Douglas, *Purity and Danger: An Analysis of Concepts of Pollution and Taboo*, London 1966, 1985.

5 Robert Jütte, *Ärzte, Heiler und Patienten: Medizinischer Alltag in der frühen Neuzeit*, Munich 1991.

6 Josef Ehmer, 'Demographische Krisen, demographische Transition', in *Enzyklopädie der Neuzeit* 2 (2005), columns 899–914.

7 Karl Heinz Ludwig and Volker Schmidtchen, *Metalle und Macht, 1000–1600* [= *Propyläen Technikgeschichte*, vol. 2], Berlin 1992, pp. 76–106.

8 Paul Mantoux, *The Industrial Revolution in the Eighteenth Century*, London 1928, p. 224.

9 Mark Overton, *Agricultural Revolution in England*, Cambridge 1996.

10 David S. Landes, *The Unbound Prometheus*, Cambridge 1969, pp. 89ff.

11 Paul Mantoux, *The Industrial Revolution in the Eighteenth Century*, London 1928, pp. 189–219.

12 Akos Paulinyi and Ulrich Troitzsch, *Mechanisierung und Maschinisierung 1600 bis 1840* [= *Propyläen Technikgeschichte*, vol. 3], Berlin 1991.

13 Felix Butschek, *Europa und die Industrielle Revolution*, Vienna 2002.

14 Paulinyi and Troitzsch, *Mechanisierung und Maschinisierung*, pp. 353–68.

15 David S. Landes, *The Unbound Prometheus*, Cambridge 1969, pp. 93f.

16 Carlo Cipolla (ed.), *The Industrial Revolution, 1700–1914*, Hassocks, Sussex 1976, p. 4.

17 Jörn Sieglerschmidt (ed.), *Der Aufbruch ins Schlaraffenland: Stellen die Fünfziger Jahre eine Epochenschwelle im Mensch-Umwelt Verhältnis dar?* [= *Environmental History Newsletter* 2], Mannheim 1995.

18 Christian Pfister, 'Das 1950er Syndrom – die umweltgeschichtliche Epochenschwelle zwischen Industriegesellschaft und Konsumgesellschaft', in Sieglerschmidt (ed.), *Der Aufbruch ins Schlaraffenland*, pp. 28–71.

19 Rolf Peter Sieferle, *The Subterranean Forest: Energy Systems and the Industrial Revolution*, Cambridge 2001.

20 Donella and Dennis Meadows, *Limits to Growth. The 30-Year Update*, White River Jct., VT 2004.

21 Carlo M. Cipolla, *The Economic History of World Population*, New York 1978, pp. 113–17.

22 'Bevölkerungsentwicklung', in *Brockhaus Enzyklopädie*, vol. 3 (2006), pp. 789–94.

23 Rolf Peter Sieferle, *Fortschrittsfeinde? Opposition gegen Technik und Industrie von der Romantik bis zur Gegenwart*, Munich 1984.

24 Ulrich Linse, *'Barfüßige Propheten': Erlöser der zwanziger Jahre*, Berlin 1983.

25 'Club of Rome', in *Brockhaus Enzyklopädie*, vol. 5 (2006), pp. 761–2.

26 Donella H. Meadows et al., *The Limits to Growth: A Report for the Club of Rome's Project on the Predicament of Mankind*, London 1974, p. 183.

27 Silvia Liebrich, 'Grönland hofft auf Ölreichtum', *Süddeutsche Zeitung*, 25 July 2006, p. 26.

28 Spencer R. Weart, *The Discovery of Global Warming*, Cambridge, MA 2003, pp. 3f.

29 Svante Arrhenius, 'On the Influence of Carbonic Acid in the Air upon the Temperature of the Ground', *Philosophical Magazine* 41 (1896), pp. 237–76.

30 Henning Rohde and Robert Charlson (eds), *The Legacy of Svante Arrhenius*, Stockholm 1998.

31 Weart, *The Discovery of Global Warming*, pp. 6f.

32 Guy Stewart Callendar, 'The Artificial Production of Carbon Dioxide and Its Influence on Temperature', *Quarterly Journal of the Royal Meteorological Society* 64 (1938), pp. 223–40.

33 Gilbert N. Plass, 'The Carbon Dioxide Theory of Climatic Change', *Tellus* 8 (1956), pp. 140–54.

34 James Rodger Fleming, *Historical Perspectives in Climate Change*, New York 1998, diag. 9–5.

35 Stefan Rahmstorf and Hans Joachim Schellnhuber, *Der Klimawandel*, Munich 2006, p. 33.

36 Weart, *The Discovery of Global Warming*, pp. 68f.

37 Paul Andrew Mayewski and Frank White, *The Ice Chronicles*, London 2002, pp. 24f.

38 J. Murray Mitchell Jr., 'Recent Secular Changes of Global Temperature', *Annals of the New York Academy of Sciences* 95 (1961), pp. 247–9.

39 George J. Kukla and R. K. Matthews, 'When Will the Present Interglacial End?', *Science* 178 (1972), pp. 190–1.

40 A. C. Stern (ed.), *Air Pollution*, 2 vols., New York 1962.

41 Reid A. Bryson and Wayne M. Wendland, 'Climatic Effects of Atmospheric Pollution', in S. Fred Singer (ed.), *Global Effects of Environmental Pollution*, New York 1970, pp. 130–8.

42 J. Murray Mitchell Jr., 'A Preliminary Evaluation of Atmospheric Pollution as a Cause of the Global Temperature Fluctuation of the Past Century', in Singer (ed.), *Global Effects of Environmental Pollution*, pp. 139–55.

43 Tor Bergeron, 'Richtlinien einer dynamischen Klimatologie', *Meteorologische Zeitschrift* 47 (1930), pp. 246–62.

44 Hubert H. Lamb, *Climate History and the Modern World*, London 1982, pp. 14f., 365, 376.

45 Lowell Ponte, *The Cooling*, Englewood Cliffs, NJ 1976, pp. 217–33, 239f.

46 Syukuro Manabe and R. T. Wetherald, 'Thermal Equilibrium of the Atmosphere with a Given Distribution of Relative Humidity', *Journal of Atmospheric Science* 24 (1967), pp. 241–59.

47 Syukuro Manabe, 'The Dependence of Atmospheric Temperature on the Concentration of Carbon Dioxide', in Singer (ed.), *Global Effects of Environmental Pollution*, pp. 25–9.

48 Wallace S. Broecker, 'Are We on the Brink of a Pronounced Global Warming?', *Science* 189 (1975), pp. 460–4.

49 Stephen H. Schneider, 'Editorial for the First Issue of Climate Change', *Climate Change* 1 (1977), pp. 3–4.

50 W. W. Kellogg and R. Schware, *Climate Change and Society*, Boulder, CO 1981.

51 R. E. Dickinson and R. J. Cicerone, 'Future Global Warming from Atmospheric Trace Gases', *Nature* 319 (1986), pp. 109–15.

52 G. I. Pearman (ed.), *Greenhouse: Planning for Climate Change*, New York 1988.

53 T. J. Marsh, R. A. Monkhouse, N. Arnell et al., *The 1988–92 Drought*, Wallingford 1994.

54 Stephen H. Schneider, *Global Warming*, New York 1990, 'Preface' and pp. 13ff.

55 Robert Lichter, *A Study of National Media Coverage of Global Climate Change 1985–1991*, Washington, DC 1992.

56 G. Stanhill, 'The Growth of Climate Change Science: A Scientometric Study', *Climatic Change* 48 (2001), pp. 515–24.

57 Karl Heinz Ludwig, *Eine kurze Geschichte des Klimas*, Munich 2006, pp. 136ff., 154ff.

58 John T. Houghton et al. (eds), *Climate Change: The IPCC Scientific Assessment*, Cambridge 1990.

59 Albert Gore, *Earth in Balance: Ecology and the Human Spirit*, Boston 1992.

60 Roger Bate and Julian Morris, *Global Warming: Apocalypse or Hot Air?*, London 1994.

61 John T. Houghton et al., *Climate Change: The Science of Climate Change*, Cambridge 1995.

62 Ludwig, *Eine kurze Geschichte des Klimas*, pp. 156ff.

63 Tim Flannery, *The Weather Makers*, London 2006, pp. 199ff.

64 IPCC (ed.), *Special Report on Emission Scenarios: A Special Report of Working Group III of the IPCC*, Cambridge 2000.

65 Stefan Rahmstorf and Hans Joachim Schellnhuber, *Der Klimawandel*, Munich 2006, p. 49.

66 John T. Houghton et al. (eds), *Climate Change 2001: The Scientific Basis. Contribution of Working Group I to the Third Assessment Report of the IPCC*, Cambridge 2001.

67 Dick Tavern, 'Vergesst Kyoto!', *Frankfurter Allgemeine Sonntagszeitung*, 28 August 2005, p. 37.

68 Karl Heinz Ludwig, *Eine kurze Geschichte des Klimas*, pp. 168ff.

69 *Süddeutsche Zeitung*, 3–4 February 2007, pp. 1, 2, 4; *Frankfurter Allgemeine Zeitung*, 3–4 February 2007, pp. 1, 2, 33.

70 Richard Alley et al., *IPCC Climate Change 2007: The Physical Sciences Basis – Summary for Policymakers*, 2 February 2007.

71 Christian Schwägerl, 'Wir müssen das Fossilzeitalter beenden', *Frankfurter Allgemeine Zeitung*, 3 February 2007, p. 33.

72 Camille Parmesan and Gary Yohe, 'A Globally Coherent Fingerprint of Climate Change Impacts across Natural Systems', *Nature* 421 (2003), pp. 37–42.

73 Camille Parmesan et al., 'Poleward Shifts in Geographical Ranges of Butterfly Species Associated with Global Warming', *Nature* 399 (1999), pp. 579–84.

74 Heiko Lehmann, 'Flinker Falter aus dem Süden: Taubenschwänzchen wird bei uns heimisch', *Saarbrücker Zeitung*, 19–20 August 2006, p. B6.

75 T. L. Root et al., 'Fingerprints of Global Warming on Wild Animals and Plants', *Nature* 421 (2003), pp. 57–60.

76 Peter Boehm, 'Global Warming – Devastation of an Atoll', *The Independent*, 30 August 2006, pp. 24–5.

77 Joe Barnett and Neil Adger, 'Climate Dangers and Atoll Countries', *Climatic Change* 61 (2003), pp. 321–37.

78 Andrea Bigano et al., 'The Impact of Climate on Holiday Destination Choice', *Climatic Change* 76 (2006), pp. 389–406.

79 J. Jason West et al., 'Storms, Investor Decisions, and the Economic Impacts of Sea Level Rise', *Climatic Change* 48 (2001), pp. 317–42.

80 R. McLeman and B. Smit, 'Migration as an Adaptation to Climate Change', *Climatic Change* 76 (2006), pp. 31–53.

81 Joachim Müller-Jung, 'Schleusen auf und weg damit', *Frankfurter Allgemeine Zeitung*, 9 August 2006, p. 31.

82 Christiane Grefe, ' "Die Reichen sollen bezahlen" ', *Die Zeit* 33, 10 August 2006, p. 19.

83 Joachim Müller-Jung, 'Schleusen auf und weg damit', p. 31.

84 IPCC (ed.), *Special Report on Carbon Dioxide Capture and Storage*, Geneva 2005.

85 Stefan Rahmstorf and Hans Joachim Schellnhuber, *Der Klimawandel*, Munich 2006, pp. 109ff.

86 Paul J. Crutzen, 'Albedo Enhancement by Stratospheric Sulfur Injection: A Contribution to Resolve a Policy Dilemma?', *Climatic Change* 77 (2006), pp. 211–20.

87 Rahmstorf and Schellnhuber, *Der Klimawandel*, pp. 133f.

88 Müller-Jung, 'Schleusen auf und weg damit', p. 31.

6 EPILOGUE: SINS AGAINST THE ENVIRONMENT AND GREENHOUSE CLIMATE

1 Robert F. Kennedy, *Crimes against Nature*, New York 2004.

2 Robert F. Kennedy Jr, 'For They That Sow the Wind Shall Reap the Whirlwind', posted 29 August 2005, http://www.stopglo balwarming.org.

3   Richard B. Alley, 'Temperatursprünge: Das instabile Klima', in *Spektrum der Wissenschaft: Dossier 2* (2005), pp. 6–13: here, p. 8. Cf. idem, 'Abrupt Climate Change', *Scientific American*, November 2004, p. 64: 'Scientists cannot yet predict when such abrupt changes will occur, but most climate experts warn that global warming and human activities may be propelling the world faster toward sudden, long-lasting climate changes.'

4   Andreas Mihailescu, *Umweltsünden-Katalog*, Munich 1983.

5   James E. Hansen, 'Earth's Energy Imbalance: Confirmation and Implications', *Science* 308 (2005), pp. 1431–5.

6   Richard C. J. Somerville, 'Medical Metaphors for Climate Issues', *Climatic Change* 76 (2006), pp. 1–6.

7   Harald Kohl and Helmut Kühr, 'Treibhausszenario: Klimawandel der Erde – die planetare Krankheit', in *Spektrum der Wissenschaft: Dossier 2* (2005), pp. 24–31: here, p. 28.

8   *Spektrum der Wissenschaft: Dossier 2, Die Erde im Treibhaus* (2005), p. 30.

9   Paul J. Crutzen et al., 'The Anthropocene', *IGBP Newsletter* 41 (2000), p. 12.

10 William F. Ruddiman, 'The Anthropogenic Greenhouse Era Began Thousands of Years Ago', *Climatic Change* 61 (2003), pp. 261–93.

11 *Brockhaus Enzyklopädie*, vol. 3 (2006), p. 790.

12 Paul J. Crutzen and Will Steffen, 'How Long Have We Been in the Anthropocene Era?', *Climatic Change* 61 (2003), pp. 251–7.

13 Thomas J. Crowley, 'When Did Global Warming Start?', *Climatic Change* 61 (2003), pp. 259–60.

14 Martin Claussen et al., 'Did Humankind Prevent a Holocene Glaciation? Comment on Ruddiman's Hypothesis of a Pre-Historic Anthropocene', *Climatic Change* 69 (2005), pp. 409–17.

15 William F. Ruddiman, 'The Early Anthropocenic Hypothesis a Year Later', *Climatic Change* 69 (2005), pp. 427–34.

16 Stefan Rahmstorf and Hans Joachim Schellnhuber, *Der Klimawandel*, Munich 2006, p. 124.

17 Naomi Oreskes, 'The Scientific Consensus on Climate', *Science* 306 (2004), p. 1686.

18 James Lovelock, *Gaia: A New Look at Life on Earth*, Oxford 1979.

19 Richard Alley, *The Two-Mile Time Machine*, Princeton, NJ 2002, p. 4.

20 Spencer R. Weart, *The Discovery of Global Warming*, Cambridge, MA 2003, pp. 199ff.

21 Wissenschaftlicher Beirat der Bundesregierung Globale Umweltveränderungen (ed.), *Welt im Wandel – Energiewende zur Nachhaltigkeit*, Berlin 2003.

22 Arnold Toynbee, *A Study of History*, 12 vols., London 1934–61.

23 Karl Heinz Ludwig, *Eine kurze Geschichte des Klimas*, Munich 2006.

24 Harald Martenstein, 'Hamburg: ein Phänomen wie die Kalahariwüste', *Geo kompakt: Wetter und Klima* (2006), pp. 152–3.

25 Joachim Radkau, *Natur und Macht: eine Weltgeschichte der Umwelt*, Munich 2000, p. 48.

26 For example: Thomas S. Kuhn, *The Structure of Scientific Revolutions*, Chicago 1962.

27 Augusto Mangini, 'Ihr kennt die wahren Gründe nicht', *Frankfurter Allgemeine Zeitung*, 5 April 2007, p. 35.

FURTHER READING

Trevor Aston (ed.), *Crisis in Europe 1560–1660*, London 1965.

Wolfgang Behringer, Hartmut Lehmann and Christian Pfister (eds), *Kulturelle Konsequenzen der Kleinen Eiszeit*, Göttingen 2004.

Wolfgang Behringer, *Witches and Witch Hunts: A Global History*, Cambridge 2004.

Tim Flannery, *The Weather Makers: The History and Future Impact of Climate Change*, London 2006.

Rüdiger Glaser, *Klimageschichte Mitteleuropas*, Darmstadt 2001.

František Graus, *Pest – Geißler – Judenmorde: Das 14. Jahrhundert als Krisenzeit*, Göttingen 1987.

Jean M. Grove, *The Little Ice Age*, London 1988.

John T. Houghton (ed.), *Global Warming: The Complete Briefing*, Cambridge 1997.

Manfred Jakubowski-Tiessen/Hartmut Lehmann (eds), *Um Himmels Willen: Religion in Katastrophenzeiten*, Göttingen 2003.

William Chester Jordan, *The Great Famine: Northern Europe in the Early Fourteenth Century*, Princeton 1996.

Emmanuel Le Roy Ladurie, *Times of Feast, Times of Famine: A History of Climate since the year 1000*, London 1972.

Davis S. Landes, *The Unbound Prometheus: Technological Change and Industrial Development in Western Europe from 1750 to the Present*, London 1969.

Wilhelm Lauer and Jörg Bendix, *Klimatologie*, Brunswick 2004.

Paul Andrew Mayewski and Frank White, *The Ice Chronicles*, London 2002.

Stephen Mithen, *After the Ice: A Global Human History, 20000–5000 BC*, London 2003.

Christian Pfister, *Klimageschichte der Schweiz 1525–1860*, Berne 1988.

Theodore K. Rabb, *The Struggle for Stability in Early Modern Europe*, New York 1975.

Joachim Radkau, *Natur und Macht: Eine Weltgeschichte der Umwelt*, Munich 2000.

Stefan Rahmstorf and Hans Joachim Schellnhuber, *Der Klimawandel*, Munich 2006.

John E. Richards, *The Unending Frontier: An Environmental History of the Early Modern World*, Berkeley 2003.

Stephen H. Schneider and Randi Londer, *The Coevolution of Climate and Life*, San Francisco 1984.

Steven Shapin, *The Scientific Revolution*, Chicago 1996.

Hansjörg Siegenthaler, *Regelvertrauen, Prosperität und Krisen: Die Ungleichmäßigkeit wirtschaftlicher und sozialer Entwicklung als Ergebnis individuellen Handelns und sozialen Lernens*, Tübingen 1993.

Tom Simkin and Lee Siebert, *Volcanoes of the World*, Tucson, AZ 1994.

Steven M. Stanley, *Earth and Life through Time*, 2nd edn, New York 1989.

Nico Stehr and Hans von Storch, *Klima, Wetter, Mensch*, Munich 1999.

Spencer R. Weart, *The Discovery of Global Warming*, Cambridge, MA 2003.

INDEX

NOTE: Page numbers followed by *fig* refer to a figure; there may also be relevant information in the text.